Humic Substances

Edited by Abdelhadi Makan

Published in London, United Kingdom

IntechOpen

Supporting open minds since 2005

Humic Substances
http://dx.doi.org/10.5772/intechopen.92966
Edited by Abdelhadi Makan

Contributors
Priya Goel, Madhu Dhingra, Alabi Olusoji David, Juliano de Oliveira De Oliveira Barbirato, Natália C. Ferreira, Leonardo B. Dobbss, Inna Vladimirovna Simakova, Aleksey Alekseevich Vasiliev, Konstantin Vyacheslavovich Korsakov, Vladimir Vasilievich Salautin, Lyudmila Yurievna Gulyaeva, Nikita Olegovich Dmitriev, Rajneesh Kumar Gautam, Islamuddin, Dimuth Navaratna, Shobha Muthukumaran, Amarendra Singh, Nandkishor More, Claudio Fernando Mahler, Nicoly Dal Santo Svierzoski, Cassiano Augusto Rolim Bernardino, Pilar Mañas, Jorge De Las Heras, Hemlata K. Bagla, Abdennaceur Hassen, wafa hassen, Bilel Hassen, Rim Werheni, Yassine Hidri, Lyudmila Alexandrovna Sivokhina

First published in London, United Kingdom, 2021 by IntechOpen
IntechOpen is the global imprint of INTECHOPEN LIMITED, registered in England and Wales, registration number: 11086078, 5 Princes Gate Court, London, SW7 2QJ, United Kingdom
Printed in Croatia

British Library Cataloguing-in-Publication Data
A catalogue record for this book is available from the British Library

Additional hard and PDF copies can be obtained from orders@intechopen.com

Humic Substances
Edited by Abdelhadi Makan
p. cm.
Print ISBN 978-1-83962-751-4
Online ISBN 978-1-83962-752-1
eBook (PDF) ISBN 978-1-83962-753-8

For EU product safety concerns:
IN TECH d.o.o., Prolaz Marije Krucifikse Kozulić 3, 51000 Rijeka, Croatia, info@intechopen.com or visit our website at intechopen.com.

Meet the editor

Abdelhadi Makan has a Ph.D. in Waste Management and Environmental Sciences. He gained relevant professional experience in fields such as environmental science, chemistry, and waste management. Through all his academic and professional publications, he has fully practiced both qualitative and quantitative research methods. Additionally, he is the founder and CEO of ENQUAS Consulting, an investigation and consultancy office practicing in environmental, quality, and safety domains. Currently, he works as a full-time professor at the National School of Applied Sciences (ENSAH), Abdelmalek Essaadi University, Morocco. In his spare time, he acts as an editorial board member, reviewer, and proofreader for several reputed scientific journals.

Contents

Preface

Present in soil, water, and geological organic deposits, humic substances are the most widespread organic compounds, naturally occurring from the physical, chemical, and microbiological transformation of biomolecules. Humic substances represent about 25% of the total organic carbon on the Earth and comprise up to 75% of the dissolved organic carbon in water, making them important for multiple environmental processes in both soil and aquatic systems.

Based on their solubility, humic substances can be fractionated into humic acid, fulvic acid, humin, and some microelements. Humic acid includes long-chain aggregates soluble in alkali, while fulvic acid is a mixture of short-chain compounds soluble in both acidic and alkaline solutions. Humin, in contrast, constitutes the non-soluble fraction of the humate. This classification is based only on superficial criteria and does not indicate chemical behavior or provide insights into chemical structures.

Humification, or the set of processes leading to humic substances, is a set of reactions occurring either in sequences or simultaneously and affecting at varying speeds an extreme diversity of substrates. These reactions are essentially biochemical in nature, but purely chemical processes can also be involved. Physical factors, such as temperature or the humidity of the surrounding environment, also determine the formation of humic substances at various stages. Humification, therefore, appears to be a complex process in which biological activity plays a major role in the short term, but the slow evolution of humic compounds due to external conditions may gradually modify the properties of the humus.

Despite many decades of extensive studies, the formation mechanisms of humic substances, as well as their structure, are still a subject of discussion and controversy. The content of this book revolves around the structure and dynamics of humic substances, their physicochemical and biological properties, and the technics of their extraction and characterization. The book also sheds light on recent advances and applications of humic substances in agriculture, environment, industry, and medicine.

Abdelhadi Makan
Abdelmalek Essaadi University,
Tétouan, Morocco

Chapter 1

Humic Substances: Prospects for Use in Agriculture and Medicine

Priya Goel and Madhu Dhingra

Abstract

Humic substances are the naturally occurring biogenic, heterogeneous organic compounds resulting from the decomposition of plant, animal and microbial residues, over many thousands of years. Research into humic substances have found them to be beneficial not only for crop production, but for human and animal health also. Humic substances influence crop yields by affecting soil texture, water retention, nutrient buffering capacity and microbial diversity. Their stimulatory effects on plant growth and physiology can be ascribed to enhanced nutrient uptake, hormonal activity with root growth and proliferation, activation of antioxidant defense under various abiotic stresses. Furthermore, their potential to remove toxic pollutants from soil as well as aquatic systems might resolve the problem of degrading environment. Positive effects of humic substance on soil as well as crop without harming the environment make humic acid a sustainable solution for crop production. The antioxidant, immunomodulatory, anti-inflammatory, detoxifying and nutritional properties of humic substances opens up new avenues for their medical applications. Still, more research is needed to completely explore their potential. This chapter will summarize and review information about the prospects of using humic substances in agriculture and medicine.

Keywords: Humic substances, Fulvic acid, Humic acid, agriculture, medicine

1. Introduction

Humic substances (HS) are soil-derived substances that form one of the vast reservoirs of organic carbon in nature. HS are primarily produced by physical, chemical and microbial degradation and transformation of plant and animal tissues (humification process) over millions of years. They constitute the bulk of organic matter of the soil – humus, peat, lignites and also brown coal. HSs may also be produced as a by-product of the synthetic oxidation reactions of phenolic compounds.

Humic substances contain carbon, hydrogen, oxygen and nitrogen with small amounts of sulfur and phosphorus. They are a mixture of acids that can be fractionated on the basis of differences in their solubility into humic acid (HA), fulvic acid (FvA) and humin fractions, beside ulmic acid, and some microelements. Humic acid includes aggregates of long chain, high molecular weight compounds, soluble in alkali, while FVA is a mixture of short chain, low molecular weight compounds, soluble in both acidic and alkaline solutions. Humin, in contrast, constitutes the non-soluble fraction of the humate.

Owing to their unique structural and chemical properties, humic substances make an excellent soil amendment. They offer beneficial impacts in terms of overall soil structure, water holding capacity and nutrient availability in the soil along with

IntechOpen

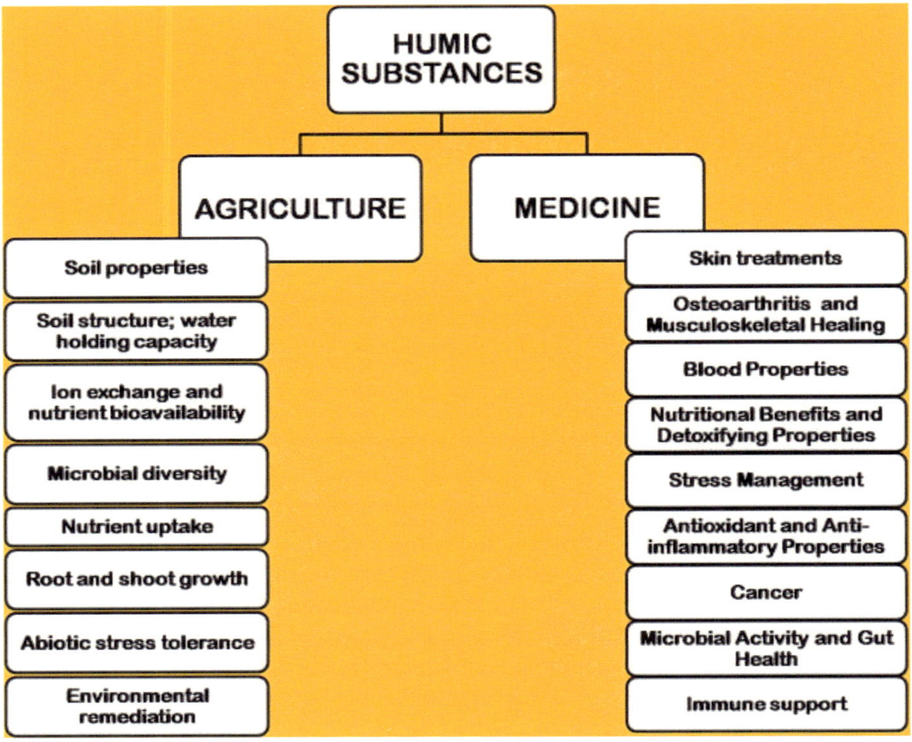

Figure 1.
Applications of Humic Substances in various fields of agriculture and medicine.

positive effect on the microbial communities in the soil, hormone like impact on plant growth. Thus, proving their vitality for farmers' economic status. HSs have also been applied in the treatment of various human diseases and maintaining overall vigour and growth of the body. Being rich in minerals which are easily assimilated in our body, their consumption in the form of dietary supplements and addition in cosmetic products is highly recommended. **Figure 1** tabulates various fields of agriculture and medicine where humic substances have shown their applications.

2. Role of humic substances in agriculture

2.1 Impact on soil properties

Humic substances play an important role with their effect on the quality and productivity of the soil. They influence the soil structure and various physico-chemical properties and are involved in the majority of soil surface phenomena. Humic substances, increase soil aggregation, water retention, infiltration rate, and water-holding capacity [1]. Research shows that aggregation in soils is improved with humic substances over a wide range of texture grades and mineral suites [2]. Polymerization reaction between the hydroxyl group and carboxyl group of humic acid in the fertilizer with the calcium in the soil, lessens the soil bulk density, improves the porosity, provides good permeability; thus, improving the soil structure [3]. Significant differences have been observed between the field capacity, permanent wilting point, and available water capacity of untreated and humic acid treated soils and emphasized that humic substances can improve structure of degraded soils even [4]. The combination of colloid in humic acid and the calcium in the soil forms flocculent gels that impacts the soil loosening, increases air permeability and water storage capacity and develops soil aggregate structure [5]. Humic acids have potential to improve soil structure in short term as aggregate stability increased with humic acid application while soil modulus of rupture decreased [6]. The application of humic acids in arid white sandy and red sticky soil can provide the opportunity for use of reserved cultivated land resources, as these substances decrease the bulk density band increase the porosity of soil [7]. Positive effect have been observed of humic acid application on physico-chemical characteristics of soil embankment with pH value in range of 7–8, cation exchange to 60% saturated and soil bulk density in the range of 1.1–0.97 g/cm^3, thus making embankment favorable for fish or shrimp [8].

Humic acid can improve the chemical properties of the soil. The application of humic substances influences the cation exchange capacity of the soil. Exogenous application of humic acid increased exchangeable Na, Ca, Mg and K while decreased soil pH and EC [9]. Similar positive effects on soil EC, organic matter and total nitrogen were reported with increasing dose of humic acid [6]. Humic substances buffer the soil pH and release carbon dioxide. The biological and physiochemical properties of organic substances may play role in nitrogen mineralization from organic sources during incubation [10]. Long term effect of humic acid application on soil nutrient status was confirmed through a three-year experiment on peanut in which humic acid treatment increased not only soil total nitrogen, phosphorus, potassium content and available NPK contents but also the organic matter content [11]. Increase in organic matter content of the soil can be useful for the improvement of agricultural soils [12]. In saline-alkali soils, humic acid adsorbs soluble salts in the soil, obstructs unfavorable cations, and decreases the salt concentration and pH of the soil. These substances have potential to counterbalance the acidity and salinity of the soil and act as a natural chelator for metal ions under alkaline conditions. It exchanges H^+ ions with Na in the soil thereby declining Na^+ content and increasing H^+ levels, consequently soil pH and SAR is reduced [9]. Humic substances decrease the soil Na, EC and pH likely due to high supplies of Ca, Mg and K which hold the cation-exchange sites on soil particles, restrict the Na adsorption and thus enhance Na leaching losses during precipitation events [13].

Soil microorganisms are important components involved in the processes viz., decomposition of soil organic matter, humus formation and soil nutrient cycling [14, 15]. Humic substances affect the structure of microbial community and impact the soil fertility [16]. The activity of soil enzymes represents the metabolism in the soil, which in turn reflects the nutrient uptake and growth of plants. Humic acid treatment increased the activities of soil enzymes- urease, sucrose and phosphates in a three year experiment on peanut [11]. Furthermore, microbial diversity of the soil was also altered with the application of humic acid; promotion of bacteria while the effect was reverse with fungi. Humic acids facilitate the growth of microbes by providing a source of carbon and food for micro-organisms. They also provide a source for colonization of microorganisms because of their large size; thus, providing continuous supply of nutrients for establishment of beneficial microbial communities in soil [17]. Humic acids at a concentration of 0.1 g/l stimulated the growth of the soil bacteria, probably, modulating the cell metabolism [18]. Application of humic substances has significant impact on carbon and nitrogen transformation processes occurring in the soil. This effect may be attributed to the functional groups of humic substances, which can be readily oxidized and subsequently act as electron donors for bacterial respiration, thus affecting the CO_2 and CH_4 production [19]. Microbial processes affected by humic acids can help to reduce greenhouse gas emissions [20].

2.2 Impact on plant growth

Humic substances are fairly stable products of organic matter decomposition and, thus, accumulate in the environmental systems. Due to their advantageous effects on different soil properties, their role in plant growth has been recognized. For several decades, commercial humic products have been available and applied to increase crop growth and economic yield [21]. Commercially available humic substance-based products can be applied both to the soil (root area) as well as foliar sprays. The impact of soil application has been studied extensively but for foliar application mechanism needs to be explored. In foliar application, humic substances do not come in contact with rhizosphere where important effect of humic substances on nutrient availability takes place [22]. Two hypothesis have been proposed to explain the effect of humic substances on plant growth- (a) Nutritional hypothesis, which proposes that the humic substances affect the plants due to their indirect effect on soil properties and the mineral nutrients available in the soil and (b) Hormone-like hypothesis, which proposes their direct impact on plant metabolism through roots [23]. The impact of humic acids on the plants depends on the concentration used, their genesis, molecular size and number and distribution of hydrophilic and hydrophobic sites. Humic substances mainly act upon proteins involved in nutrient transport, nitrogen assimilation, cell division and development; plasma membrane H^+-ATPases, hormone pathways [24].

2.2.1 Nutrient uptake from soil

Because of their structure and chemical composition, humic acids make nutrients more available in soil that might not otherwise be available. Negatively charged humic substances readily attract the cations and bind with them; thus affecting the cation exchange capacity of the soil. Humic substances form complexes with metallic ions and enhance the availability of micronutrients like Zn, Mn, Cu, and Fe and macronutrients like P, and specifically in the soils deficient in these nutrients [25]. Liquid fulvic acids polarize the soil and increase the uptake of nutrients [26]. So, humic substances form systems for ion exchange and complexation of metals, thus exerting beneficial effects on nutrient uptake, their availability and transport. Many researchers reported positive effect of both soil and foliar application of humic acids on uptake of different macronutrients (N, Mg, Ca, K and P) and micronutrients (Cu, Zn, Mn, Fe) in various crops viz., lettuce [27], maize [28, 29], wheat [30], peppers and cucumbers [16].

Application of either dry or liquid humic substances to soils improves the fertilizer efficiency. Humic substances diminish the leaching of nitrogen compounds into subsoil water and reduce the toxicity. They bind to these nutrients in a molecular form which decreases their solubility in water; thus, increasing soil exchangeable NH^{4+} and available NO^{3-} while retaining more nitrogen in the soil. These binding processes also prevent volatilization of N into the atmosphere. This role of humic substances is related to their effects on plant cell membranes, and associated biochemical and molecular processes initiated at post-transcriptional level in roots and shoots [31]. Humic substances might increase the uptake efficiency of nutrients, because of their ability to alter the activity of the root membrane H^+-ATPase, which is a recognized as a marker of biostimulant action and contributes to regulate the rhizosphere pH, thereby affecting nutrient availability [32].

2.2.2 Root and shoot growth

Humic compounds have been reported to improve plant growth in terms of parameters like plant height and weight. These effects vary with the concentration and origin of the humic substance and the plant species. Applications of humic and fulvic acids at lower concentrations is advantageous for root proliferation. However, use of these compounds at high concentrations might decrease root and shoot growth. The positive effects of humic substances on root and shoot growth can be ascribed to their hormone like activities as different hormones have been identified in the humus structure [33]. The potential capacity of humic substances to induce root development in plants with auxin-like activity was first postulated by an independent research group working on mobilization of soil organic matter by root exudates [34].

Humic acids prevent the oxidation and breakdown of hormones, thus, auxins remain active for longer time. Application of humic acid, increase the root and thus, the plant is better able to take up water and nutrients. Auxins induce activity of plasma membrane H^+-ATPase root cells, which couple ATP hydrolysis to H^+ transport across cell membranes. The apoplast is acidified, the cell walls are loosened and resulting in cell elongation, thereby increasing the root growth. The stimulation of H^+-ATPase affects nutrient uptake, by enhancing the electrochemical proton gradient that drives ion transport across cell membranes. Humic acid-like auxins also induce H^+-ATPase synthesis and it activity [35]. Humic substances induce the synthesis of plasma membrane and increase the root growth with similar mechanism. Humic acid treated maize seedlings recorded higher root fresh weight than control plants along with increased number, diameter and length of roots [36]. Similar effects of root elongation, lateral root formation and increased activity of H^+-ATPase were observed in maize seedlings treated with humic substances obtained from earthworm compost which showed the presence of exchangeable auxin groups in the structural analysis [37]. Mechanism of effect of humic substances on root structure is still not clear. HA is also hypothesized to affect root architecture in cucumber seedlings and enhance nitric oxide, ethylene and auxin levels though the latter not being responsible for the former [38]. He suggested the participation of some other factors in improvement of root architecture with humic acid application. Also, research has shown that foliar application of humic substances did not result in increased activity of H^+-ATPase and various hormones like ABA as observed with application to the root; while both treatments increased the jasmonic acid content in root which can be related to adaptation to mild transient stress caused by humic substance application [22].

Humic substances also affect shoot growth, and it has been shown to affect multiple plant processes like photosynthesis, respiration, protein metabolism and activities of enzymes, uptake of water and nutrients. The mechanisms for these effects have been related to changes in hormones, cell membrane permeability, components of ETC, hydroxyproline:proline ratio, free radical activity within the humic structure, and reactive oxygen species in plants [21]. Although it is still not fully understood, humic acids result in better shoot growth. Application of humic acidsimproves the health and vigor of the plant. Apart from auxin, humic substances also exhibit gibberellins and cytokinin like activities [39]. These activities stimulate plant metabolism and in turn shoot growth. It has also been speculated that effect of humic substances on H^+-ATPase might affect the root-shoot distribution of various hormones like cytokinins, polyamines etc., thus improving the shoot growth [40]. Foliar applications of humic acid resulted in higher leaf and stem dry matter contents than the control in tomato [41]. Hormone like activity of humic acids also increased the number of harvested flowers per plant and extended the vase life of harvested flowers in gerbera [42]. Humic acid treatment showed highly significant effect on dry biomass, root/shoot ratio and photosynthesis parameters like stomatal conductance, transpiration rate, photosynthetic rate in potato along with 23–37% increase in tuber yield [43].

2.2.3 Abiotic stress tolerance

The effect of humic substances in the alleviation of abiotic stress effects in plants is ascribed to the elevation in enzymatic and non-enzymatic antioxidants, formation of compatible solutes and changes in ionic balance. In various abiotic stresses, reactive oxygen species (ROS) are one of the significant causes responsible for cellular damage. The oxidative damage due to ROS is prevented by induction of enzymatic antioxidants such as superoxide dismutase, peroxidases, catalase and ascorbate peroxidase and non-enzymatic antioxidants like ascorbate, tocopherol and phenolics and production of osmo-protectants like proline, sugars. These stress combating mechanisms are set off by humic substances also. Response of humic acid treatment has been evaluated on tolerant and sensitive melon varieties under drought and reported increased shoot weight, leaf area, chlorophyll content and activities of antioxidant enzymes with humic acid application under stress [44]. Humic acids enhance the stress tolerance by reducing the evaporation of water with reducing effect on the stomatal opening; helping the plant and soil to retain more moisture [7]. The oxidative stress is prevented by humic substances through their positive effect on peroxidase activity and proline levels; thus reducing ROS levels and maintaining cellular homeostasis [45].

Humic acid treatment is also known to improve photosynthetic rate under different watering conditions through increase in rate of gas exchange and electron transport flux in rapeseed [46]. Similar effects of stimulatory effect of humic substances on proline contents, grain yield, total dry weight and harvest index was reported in chickpea under stress as well as control conditions [47]. The exploration of impact of chemical priming with humic acids on maize seedlings at transcriptional level, revealed that signaling pathways of various hormones like ABA, gibberellins and auxins and stress-responsive genes were positively modulated and humic acid priming alleviated drought, heavy metal and salinity stress effectively [48]. Application of humic acid effectively imparted tolerance to maize plants against salinity stress in terms of increased plant biomass, chlorophyll content, mineral elements and antioxidant enzymes activities [49]. An increase in permeability of root cell membranes might be responsible for enhance nutrient uptake in tomato plants grown under saline medium added with humic acid [50].

Humic substances may act through proliferation of root which is important for plant adaptation under adverse conditions like salinity for exploration of deeper layers. In addition to biochemical responses, the effects of humic substances have been studied at molecular level also. Salt-induced inhibition of HKT1 (sodium influx transporter high-affinity K^+ transporter 1) gene was overcome by application of humic acid in *Arabidopsis* through post-transcriptional control of the *HKT1* transporter gene under saline conditions [51]. Likewise, humic substances confer heat tolerance by transcriptional activation of heat shock proteins in *Arabidopsis* [52]. Thus, humic substances tend to be the potential sources to sustain crop production under arid, semi-arid and saline environments.

2.3 Environmental remediation

Humic acid-based fertilizers increase crop yield and stimulate plant enzymes and hormones in an eco-friendly manner. High adsorption capacity and high ion exchange capacity enable them to be the key candidates for environmental remediation. In environmental applications, the significant roles of humic substances are removal of toxic metals, anthropogenic organic chemicals and other chemicals like herbicides, fungicides, insecticides, etc., and pharmaceutical products from soil and water. Humic substances have the ability to adsorb, form complexes, and exert redox effects. These interact with different types of chemical groups like metal ions, hydroxides, oxides, minerals, organic matter, and toxic pollutants in the environment. In addition, these substances also participate in the biogeochemical cycles like carbon and nitrogen cycle, thus, modulating the man-made pollution in the soil. Furthermore, the detoxification organic and inorganic inhibitors of biological processes and biodegradation of toxic organic substances, speculates the efficiency of humic substances. Humic substances also hinder the uptake and availability of heavy metals *viz,* Cu, Fe, Zn etc., in soil by immobilization of these metals.

Apart from soil, humic substances in aqueous systems, affect aquatic ecosystems along with the organisms. In rivers, streams and lakes, about 50% of the dissolved organic materials are humic acids, which affect their pH; the physical and chemical properties of water are altered and they act as natural neutralizing components by stimulating biotransformation of xenobiotics. Humic acids can be deployed in water treatment and water purification technologies for removal of pollutants [53].

3. Role of humic substances in medicine

3.1 History of the use of HSs in medicine

HSs have found special place in traditional systems of medicine for roughly 3000 years in many countries, particularly India and China. Transcripts from 'Ayurveda' and 'Siddha' systems of medicine strongly refer to the use of humic substances in treatment of various ailments. Chinese Materia Medica pharmacological compendium, the medical text of the 15th century Ming Dynasty, reports the use of HSs in traditional medicine. Owing to their numerous benefits, HSs were referred to as "Wujinsan", meaning "golden medicine" in China.

The pharmacological properties of humic substances and products derived and extracted from them, have been systematically reviewed [54]. A positive approach of these substances has been documented through well-designed studies for treatment of various rheumatological and other musculoskeletal diseases. Shilajit, a traditional exudate from Himalayas in India occurs in the form of naturally occurring mineral substance. Shilajit has proven stimulatory, rejuvenatory, revitalizing, anti-ageing and anti-inflammatory properties. In addition, topical application of Shilajit has proven its antiseptic and analgesic properties [55]. Shilajit typically constitutes humic substances in which fulvic acid act is the most bioactive component (15–20%) that is known to have immunomodulatory and psychoactive behavior. Peat extracts, derived from soil – abundant in HS, have been used extensively in ancient Rome during the 19th century as mud baths for its healing effects in gynecological and rheumatic diseases [56]. Peat was also given orally to treat cardiac, gastric, intestinal or hepatic conditions.

Owing to their numerous benefits, HA and FVA have recently gained popularity in 'natural' food market. They can potentially be used as novel, natural, valuable food supplements or additives. Already some are available in the form of ready-to-drink beverages and OTC pills, but still this market needs detailed exploration in terms of combinations and variety.

3.2 Skin treatment

Direct application of mud pack on skin or as bath therapy has shown successful results in treating painful body and skin ailments. The anti-inflammatory and antioxidative properties of FvA and HA extracts present in mud help combating the symptoms and treating the condition effectively. The extracts are found safe when applied in amounts as high as 10 percent weight-by-volume. Fulvic acid or humic extracts can be applied in the form of masks, bandages, bath therapy or mud packs as in balneotherapy done in various spas round the world. These extracts are selectively absorbed via skin and have a stimulatory response on the spontaneous contractile activity of smooth muscle tissue [57].

Hospital studies have shown effective use of mud in baths or packs in curing various skin diseases, eczema, dermatitis and psoriasis including ulcers, common cold or flu, osteoarthritis and rheumatoid arthritis and other bone, joint and muscle disorders. Patients exhibit healing effects and are significantly relieved of pain and inflammation in a few sessions only. Peat is referred to as a magical, anti-ageing substance. Detailed microbial and chemical analysis of Dead Sea mineral mud, Israel used in certain dermatological and cosmetic preparations revealed that the antibacterial properties of Dead Sea mud are probably owing to some chemical and/or physical phenomena and not of the bacteria present therein [58]. Sterilization of the mud by gamma irradiation could not deter its effects. FvA and HA mineral baths have shown 90% success rate in curing ulcers; Surprisingly these baths were successful in curing internal ulcers as well [59].

3.3 Osteoarthritis and musculoskeletal healing

High temperature peat or mud bath provides depth warming, when given at specific consistency. Besides, other conditions such as temperature, ionic strength and pH of the mud or peat also influence the benefits reaped during the application. This treatment improves blood circulation and regeneration processes in the patient's body. Studies have found significant effect of mud bath therapy on osteoarthritis, a rheumatic condition associated with the progressive destruction of cartilage. Peat therapy improves the level of chondrocytes, inflammation markers- interleukin-1 (IL-1) and tumor necrosis factor alpha (TNF-α).

A trial study explained the benefits of Potassium humate treatment reducing the levels of inflammation in osteoarthritis patients over a period of 6 weeks. The levels of C-reactive protein (associated with inflammation) were significantly reduced in the blood of patients [60]. Studies on mice indicate effectiveness of Shilajit in treating induced-RA [61]. Pro-inflammatory cytokines were found to be suppressed. Its anti-arthritic mechanism of action was established through the suppression of pro-inflammatory cytokines, which suggests FVA as a potential therapeutic candidate for rheumatoid arthritis.

3.4 Blood properties

Presence of humate increases the oxygen carrying capacity of red blood cells and improves blood circulation rate. Feelings of euphoria have been reported in people taking humate in experimental studies, even during the first few days [62]. They have also shown improved rate of healing of injuries. This might be linked to the extra content of oxygen in the body. Fulvic acids have shown improved rate of absorption of iron, making higher concentrations available to bone marrow stem cells for blood formation.

Feeding animals with humic acid-rich diet has shown improvement in blood parameters viz. increased levels of total protein, globulin and glucose levels, reduced blood urea nitrogen, cholesterol, non-esterified free fatty acids and ketone body concentrations in goat fed on HA [63].

3.5 Nutritional benefits and detoxifying properties

Humic extracts, especially fulvic acids, are unique in their chelating ability. When taken orally, they provide a natural chelation therapy as an alternative medicine for removing toxins from the body. They attach to the toxins, including heavy metals, pesticides, radioactive particles, and environmental carcinogens, and hence detoxify the body - the liver and the digestive tract. The toxins are neutralized such that their cellular absorption gets difficult inside the body and get dragged out as waste products, via urine, skin, lungs or in stools. Their ability to bind toxins has led to the commercial use of humic acid in industry, animal husbandry and human nutrition. Humic acid was shown to be effective in removing botulinum neurotoxins in sublethal chronic botulism in cattle [64], removal of lead in hen thereby alleviating the toxic effects of lead poisoning on their thyroid gland [65], reduce the accumulation of lead and cadmium in fish and improving their growth rate [66].

Since they are derived from naturally occurring organic substances, humic acids particularly FvA are good source of minerals. Humic acids act as dilator, increasing the cell membrane permeability which facilitates the transfer of minerals from blood to the bone and cells. Humic substances or their extracts are medically proven to cause rehabilitation of muscles, bones and nerves and in dealing with geriatric disorders viz. arthritis, diabetes, allergic symptoms and dementia [54]. Owing to their low molecular weight and water solubility in both acidic and basic substances, FvAs are readily absorbed through semi-permeable membranes, and function actively in association with living cells. Fulvic acid acts as a chelator; it readily complexes with essential mineral ions and rare earth elements. This makes it an excellent mineral-carrier in the body. At the same time, FAs capture and remove toxic metals from the body, eliciting the process of detoxification. Within 3–4 days of FvA usage, both animal and human subjects have shown considerable results. Treatment with humic extracts have shown noticeable results in treating patients with normally incurable epidemic hemorrhagic fever. The extracts exhibited instant results in arresting bleeding, restoring circulation rates, removal of clots, beside displaying anti-viral, and immune modulatory properties.

3.6 Stress management

Humates block or reduce the production of stress causing hormones, improving the animal behavior. Humate-fed animals are less aggressive, least affected by the outside disturbances like heavy crowd or traffic or during confinement in closed arenas. Humate supplementation in diet ameliorates the adverse effects of increased caging density in hen [67, 68] suggesting the role of trace elements, having antioxidative properties, present in humates. This effect has been noted in farm animals and even fish [69].

3.7 Antioxidant and anti-inflammatory properties

Antioxidants are substances that protect cells from the damage caused by free radicals, by neutralizing their electrical charge thereby preventing the oxidative damage to the cell. Humic substances, including peat and sodium humates, are known to exhibit excellent antioxidative and anti-inflammatory properties. Fulvic acids act as good free radical scavengers, neutralizing dangerous free radicals like superoxide (O_2^-), hypochlorous acid (HOCl), hydrogen peroxide (H_2O_2), hydroxyl radicals (OH), peroxynitrite ($ONOO^-$), and singlet oxygen ($^1O2$) radicals. An increase in the level of antioxidant enzymes such as glutathione, superoxide dismutase and catalase is generally taken as a biomarker for antioxidant properties of HAs. Inside the cell, FvA can uncouple the electron transport chain in liver mitochondria, which is associated with lowering ROS production.

The ages-old FvA-rich compound, Shilajit is known to reduce hyperglycaemia in diabetic animal models and increase SOD activity in pancreatic beta cells. They have the ability to significantly increase superoxide dismutase (SOD) activity that helps preventing and combating free radical damage to pancreatic B islet cells. This property, as shown in several hospital studies, stops the advancement and progression of diabetes, plus assists in the treatment. With fulvic acid treatment, diabetes patients became more energetic and the tingling, painful feeling and numbness experienced in the nerve endings disappears or gets reduced [54].

Fulvic acid (FvA) belong to the category of natural health products (NHP) known to cure diseases associated with chronic inflammation such as diabetes, cardiovascular diseases, and colitis. They also help in relieving inflammatory states of the cervix, especially cervical erosion (generally known as cervicitis), swelling from joint inflammation as in Rheumatoid Arthritis (RH). HSs have been shown to bond to the collagen fibres to aid in repair of damaged tendons and bone.

The neuroprotective effect of HAs on cerebral ischemia rat model [70] and gerbil Hippocampus [71], therapeutic effects in renal ischemia reperfusion injury in rats [72], protective effects in iron-induced hepatotoxicity and cardiotoxicity in rats [73] have been associated with their antioxidant and free radical scavenging effects. Administration of fulvic acid to rats for 4 weeks [74] and fish for 60 days [75] showed significant decrease in lipid peroxidation and increase in the expression of antioxidant enzymes glutathione, superoxide dismutase and catalase.

3.8 Cancer research

Conventional anti-cancer therapies like chemotherapy, radiotherapy, immunotherapy and surgical approaches for cancer treatment often result in recurrence of the symptoms. During the procedures of cancer chemotherapy and radiation therapy, large amounts of free radicals are generated which damage the normal cells around the tumor cell. Hence, there has always been search for development of cancer management approaches using non-toxic and natural products.

The naturally occurring humic substances, consisting of 60–80% fulvic acids, have known anticarcinogenic properties. Naturally-occurring humic substances have proven antioxidant, anti-inflammatory, antiviral, antimutagenic, heavy metal chelating, apoptotic and photo-protective properties. These properties make them useful agents for cancer therapy and prevention. Another plus point is that when taken orally, HSs have no reported side effects and can be administered as a nutritional and rejuvenating tonic in daily routine of the patient. Numerous hospital studies have shown that humic substances, especially fulvic acids, have the power to protect against cancer and related cancer-causing viruses. Humic extracts destroy cancer cells as well as decrease cell proliferation and angiogenesis and thus, curb growth of the cancer. Special humic substance therapies have shown promising results in reversing deadly cancers and tumors.

Molecular structure of humic substances gives them excellent free radical scavenging and antioxidant properties. This antioxidant activity increases with increasing concentration of the humic extracts present in the compounds used in cancer therapy. Naturally-occurring humic substances stimulate the production of cytokines, including interferon-gamma, interferon-alpha, interferon-beta, and tumor necrosis factor-alpha.

The effect of fulvic acid on three cancer cells lines - Hep3B, HT29 and PC3 at different concentrations of FvA using MTT assay [76]. The assay revealed that FvA inhibited the proliferation of cancer cellsas well as upregulated the mRNA levels of apoptotic genes in all the three cell lines. The cytotoxic effect ofHA with Arsenic trioxide on human cervical cancer cells [77], FvA on hepatic cell lines [78] and HA on human breast adenocarcinoma cells [79] have been well documented. Even low concentrations of Mumie–a natural component found in high mountains such as the Himalayas and rich in humic compounds – could destroy human cervical cancer cells [80].

3.9 Microbial activity and gut health

Sufficient evidences link imbalance in the gut microbiome with inflammation and bowel diseases. Humic substances are known to improve the gut flora, increase the expression and activity of gut enzymes that aid in nutrient absorption and settle certain gut disorders. In other words, humic substances promote microbial activity in gut and thus, enhance nutrient absorption.

When used as feed additives, humic substances may increase rumen microbial population or shift microbiota in the gut to suit animal requirements. Increased microbial activity, in turn, improves fermentation and digestibility of nutrients in the gut rumen. HSs act as natural antibiotics, resulting in increased growth performance in farm animals and early-weaned baby animals. HSs boost the animal gut by promoting good microbes and inhibiting bad microbes when fed at high concentrations in the diet. Several studies have shown that specific species of microbes such as *Turicibacter, Clostridium, Campylobacter, Dehalobacterium, Desulfuvibrio,* and *Paludibacter* are promoted while *Prevotella, Blautia, Faecalibacterium, Lactobacillus*, and *Coprococcus* are inhibited in the cecum and colon of the animals when fed on humate supplemented diet [81, 82]. Similarly, a high relative abundance of *Serratia*, *Acinetobacter*, *Aeromonas* and *Edwardsiella* and low abundance of *Lactobacillus* was obtained in the intestine of fish fed on FvA rich diet for 60 days. The activity of digestive enzymes like lysozyme, proteases, and acid/alkaline phosphatases was also increased [75].

Humic substances particularly FvA increases the absorption of metal ions and other nutrients. There are significant proofs to show their influence on absorption of drugs in the body. FvA has also been shown to mediate drug delivery in animal models. Studies have shown that FvA enhances the uptake of iron, making it more bioavailable to bone marrow stem cells for formation of blood as well as supplying hormone stimulating micronutrients. In Broiler chicken, supplementation of diet with HA improved the utilization of nutrients and growth performance by improving the overall gut health [83, 84].

3.10 Immune support

Use of humic substances exert a beneficial effect on the animal immune system, including humans. Humates possess ability to bind sugars and assemble them to form complex saccharides within the body like glycoproteins that function as modulators of intercellular interaction. Glycoproteins bind to T cells and Killer cells, keeping them in balance and thus, regulating the immune system. Excessive T cells may lead to development of auto-immune diseases while excess killer cells can attack bone and joints causing arthritis.

Administration of HSs, especially fulvic acids, both topically and orally can bolster the immune system. They act as powerful immunomodulators, may show potent antioxidant and anti-inflammatory properties that can be used to enhance immunity and overall growth of the organism. This has been observed in growing swine [85], juvenile Loach [75] and broiler chicken [86]. Medical studies have indicated that HSs prove more effective than several currently prescribed immune regulatory drugs. Studies on mice have reported strong humoral immune stimulation by HA and FvA supplemented diets [87].

In addition, HSs have proved effective in alleviating the adverse effects of lethal doses of irradiation and reducing the incidence of infections, increasing the life span of the animal. This was directly correlated with the increasing capability of immune system to recognize its own dead cells, following administration with HSs supplemented diet or topical application of HSs. The overall health of animals improves with food and water supplemented with HSs [88].

3.11 Anti-viral properties

Anti-viral properties of soil-derived humate materials have traditionally been known to humans and have been studied experimentally since decades. Exploration into the mechanism by which humic substances inhibit viral cytopathicity revealed that these substances attach to the viral envelope protein, thereby coating its receptor sites. This further disables the virus from being able to bind and infect host cells, thus combating viral replication. Their potential range of applications cove both naked and enveloped DNA viruses.

Viral pathogens for which HSs have been found effective include Coxsackie virus A9, herpes simplex virus type 1 and 2, human immunodeficiency virus (HIV), influenza type A and B, cytomegalovirus (CMV) and Vaccinia virus as well as other respiratory tract infections. Excellent multimodal anti-HIV properties of humic substances extracted from mud [89], Shilajit [90] and Lignin-rich solid waste resulting from the processing of vegetable feedstock [91] have been demonstrated. Researches are also underway to explore the potential of HSs against the novel coronavirus COVID-19 [92].

3.12 Overall growth rate

Humic substances, with their miraculous anti-inflammatory, anti-oxidative, immunomodulatory, detoxifying, gut-promoting and nutritional properties helps in improving the overall growth and development of animal body. In animal husbandry, use of humic substances as feed additives promotes the animal growth rate, improves feed efficiency, prevents intestinal diseases and boosts overall immunity, thereby improving the economics and ecology of animal production.

Administration of HA improved the final body weight, daily weight gain and feed conversion ratio in broilers [88], increased the milk output and its quality in goats, and also growth rate of their kids with no adverse effects [63]. Even their young ones show higher survival values and better ability to overcome the effect of weaning at early stages when given diet supplemented with HSs [82].

3.13 Safety aspects

Humic substances are soil-borne materials, so humans and animals are naturally exposed to them. Industry has safely used these products for industrial exposure to heavy metals in both humans and in animals. Various studies have been conducted on both animals and humans over several years to test the effectiveness and safety of fulvic acids on human diseases [54, 93]. Most of those studies suggested that they are safe for oral or topical use, though recommended further exploration to confirm their safety to be used as food additives. No significant changes in body weights, hematological variables, indices of thyroid function, and microscopic organ histology were obtained when adult rats were orally administrated with HA [94]. No mutagenic or genotoxic effects of acute toxic and subtoxic doses of FvA in female rats [95].

However, some studies also hint upon the harmful effects HSs may be associated with. HA can induce oxidative DNA damage and genotoxicity in human lymphocytes [96]. Some studies also hint upon goitrogenic aspects of HSs [97–99] but no confirmatory studies have been made so far.

4. Conclusion

Humic substances are a group of organic macromolecules formed from dead tissues of plant and animals transformed over the years by physical, chemical and microbiological processes. They exist ubiquitously in soil, sediments and various terrestrial and aquatic environments. The entire group of HSs can be divided into three components based on their solubility: fulvic acids, humic acids, and humin fragments. Specific properties of humic substances and their products qualifies their uses in both agriculture and medicine. Exploration of compounds present in HAs and their potential in agricultural, pharmaceutical and biomedical industry is of utmost importance.

In the era of depleting natural resources, increasing food demands and degrading environment, global warming and erratic weather conditions, we cannot afford more devastation of environment by utilization of chemical fertilizers, pesticides, insecticides etc. Humic substances pose to be potential candidates for improving soil structure and nutrient status along with beneficial effects on plant growth and yield; thus, giving a pavement towards sustainable agriculture.

Primarily used to enhance growth and vigour of the organisms, HSs are increasingly being used to treat several body ailments. Their potential as anti-ageing, anti-oxidative, anti-inflammatory, anti-cancer, immunostimulatory and growth-promoting agents is largely being explored. This will help bringing these organic molecules into the niche market where they can largely replace chemical substances. So, explore and reveal the power of 'dirt'!

Author details

Priya Goel[1]* and Madhu Dhingra[2]

1 Department of Zoology, Deen Dayal Upadhyaya College (University of Delhi), Delhi, India

2 Plant Physiologist, Punjab Agricultural University, Ludhiana, Punjab, India

*Address all correspondence to: pwpriya@gmail.com

References

[1] Brady NC, RR Weil. The nature and properties of soils. Pearson Prentice Hall, Upper Saddle River, NJ. 2008

[2] Quilty J, Cattle S. Use and Understanding of Organic Amendments in Australian Agriculture: A Review. Soil Research. 2011;49(1):1-26.

[3] Baudong Fu. Research progress on the application of humic acid in soil improvement. Shelter Forest Science and Technology. 2016;03:83-84.

[4] Piccolo A, Pietramellara G. Mbagwu J. Effects of coal derived humic substances on water retention and structural stability of Mediterranean soils. Soil Use and Management. 1996;12: 209-213. https://doi.org/10.1111/j.1475-2743.1996.tb00545.x

[5] Guanglei S, Zhansheng Z, Feineng J. Effects of humic acid-containing solid waste fertilizer on spinach yield and benefits. Zhejiang Agricultural Sciences. 2016;57(11):1876-1878.

[6] Gümüş I, Seker C Influence of humic acid applications on soil physicochemical properties. Solid Earth Discuss. 2015;7:2481-2500. doi:10.5194/sed-7-2481-2015.

[7] Li Y. Research progress of humic acid fertilizer on the soil. Journal of Physics: Conference Series, 20201549 022004 doi:10.1088/1742-6596/1549/2/022004.

[8] Ali M, Mindari W (2016) Effect of humic acid on soil chemical and physical characteristics of embankment. MATEC Web of Conferences. 2016;801028.58DOI: 10.1051/conf/2016MATEC58matec501028

[9] Mindari W, Sasongko P E, Kusuma Z,Syekhfani and Aini N. Efficiency of various sources and doses of humic acid on physical and chemical properties of saline soil and growth and yield of rice. AIP Conf. Proc. 2019;030001-1-030001-8. https://doi.org/10.1063/1.5061854

[10] Yılmaz E. Changes of some soil properties by agricultural processing waste (soybean pulp) amendment. Journal of Food Agriculture and Environment. 2010;8:1057-1060.

[11] Li, Y., Fang, F., Wei, J. et al. Humic Acid Fertilizer Improved Soil Properties and Soil Microbial Diversity of Continuous Cropping Peanut: A Three-Year Experiment. Sci Rep 2019;9:12014. https://doi.org/10.1038/s41598-019-48620-4

[12] Lopez-Valdez F, Fernandez-Luqueno F, Luna-GuidoM L, MarschR, Olalde-Portugal V, Dendooven L. Microorganisms in sewage sludge added to an extreme alkaline saline soil affect carbon and nitrogen dynamics. Applied Soil Ecology. 2010;45:225-231.

[13] Lakhdar A, Rabhi M, Ghnaya T, Montemurro F, Jedidi N, Abdelly C. Effectiveness of compost use in salt-affected soil. Journal of Hazardous Materials. 2009;171:29-37.

[14] Nannipieri P, Ascher J, Ceccherini M T, Landi L, Pietramellara G, Renella G. Microbial diversity and soil functions. European Journal of Soil Science. 2003;54(4):655-670.

[15] Rajkumar M, Ae N, Prasad MN, Freitas H. Potential of siderophore-producing bacteria for improving heavy metal phytoextraction. Trends in Biotechnology. 2010;28(3):142-149.

[16] Calvo P, Nelson L, Kloepper JW. Agricultural uses of plant biostimulants. Plant and Soil. 2014;383(1):3-41.

[17] Murphy DV, Stockdale EA, Brookes PC, Goulding KWT. Impact of

microorganisms on chemical transformations in soil. In: Abbott LK, Murphy DV editors Soil Biological Fertility. Springer Netherlands, Dordrecht; 2004. p.37-59. doi: 10.1007/978-1-4020-6619-1_3

[18] Tikhonov VV, Yakushev AV, Zavgorodnyaya YA. Effects of humic acids on the growth of bacteria. Eurasian Soil Science. 2010;43:305-313. https://doi.org/10.1134/S1064229310030087

[19] Coates JD, Cole KA, Chakraborty R, O'Connor, SM, Achenbach LA. Diversity and ubiquity of bacteria capable of utilizing humic substances as electron donors for anaerobic respiration. Applied and Environmental Microbiology. 2002;68:2445-2452. doi: 10.1128/AEM.68.5.2445-2452.2002.

[20] Valenzuela EI, Cervantes FJ. The role of humic substances in mitigating greenhouse gases emissions: Current knowledge and research gaps. Science of the Total Environment. 2021;750. https://doi.org/10.1016/j.scitotenv.2020.141677.

[21] Olk DC, Dinnes D, Scoresby JR, Callaway CR, Darlington JW. Humic products in agriculture: potential benefits and research challenges—a review. Journal of Soils and Sediments. 2018. https://doi.org/10.1007/s11368-018-1916-4.

[22] De Hita D, Fuentes M, Fernández V, Zamarreño AM, Olaetxea M and García-Mina JM Discriminating the short-term action of root and foliar application of humic acids on plant growth: emerging role of jasmonic acid. Frontiers in Plant Science. 2020;11:493. doi: 10.3389/fpls.2020.00493.

[23] Garcia-Mina JM, Mora V, Olaetxea M, Baigorri R, Fuentes M, Garnica M, San Francisco S, Erro J, Urrutia O, Casanova E, Lemenager D, Yvin JC. Main mechanisms involved in the effects of humic substances on soil-plant systems. Agrociencia Uruguay. 2012;Special Issue:189-190.

[24] Nardi S, Schiavon M, Francioso O. Chemical structure and biological activity of humic substances define their role as plant growth promoters. Molecules. 2021;26:2256. https://doi.org/10.3390/molecules26082256

[25] Garcia AC, Santos LA, de Souza LGA, Tavares OCH, Zonta E, Gomes ETM. Vermicompost humic acids modulate the accumulation and metabolism of ROS in rice plants. Journal of Plant Physiology. 2016;192:56-63. DOI: 10.1155/2016/3747501.

[26] Sootahar MK, Zeng X, Wang Y, Su S, Soothar P, Bai L, Kumar M, Zhang Y, Mustafa A and Ye N. The short-term effects of mineral- and plant-derived fulvic acids on some selected soil properties: improvement in the growth, yield, and mineral nutritional status of wheat (*Triticum aestivum* l.) under soils of contrasting textures. Plants. 2020;9(2):205. doi:10.3390/plants9020205.

[27] Cimrin KM, Yilmaz I. Humic acid applications to lettuce do not improve yield but do improve phosphorus availability. Acta Agriculturae Scandinavica. 2005;55(1):58-63.

[28] Celik H, Katkat, AV, Aşık BB, Turan MA. Effect of foliar-applied humic acid to dry weight and mineral nutrient uptake of maize under calcareous soil conditions. Communications in Soil Science and Plant Analysis. 2010;42(1):29-38.

[29] Khaled H, Fawy HA. Effect of different levels of humic acids on the nutrient content, plant growth, and soil properties under conditions of salinity. Soil and Water Research. 2011;6(1):21-29.

[30] Asik BB, Turan MA, Celik H and Katkat AV. Effects of humic substanceson plant growth and mineral nutrients uptake of wheat (*Triticum durum* cv. Salihli) under conditions of salinity. Asian Journal of crop Science. 2009;1(2):87-95.

[31] Van Oosten MJ, Pepe O, De Pascale S. The role of biostimulants and bioeffectors as alleviators of abiotic stress in crop plants. Chemical and Biological Technologies in Agriculture. 2017;4, 5. https://doi.org/10.1186/s40538-017-0089-5

[32] Zandonadi DB, Santos MP, Caixeta LS, Marinho EB, Peres LE, Façanha AR. Plant protons pump as markers of biostimulant action. Scientia Agricola. 2016;73:24-28.

[33] Nardi S, Sessi E, Pizzeghello D, Sturaro A, Rella R, Parvoli G. Biological activity of soil organic matter mobilized by root exudates. Chemosphere. 2002;46:1075-1081.

[34] Concheri G, Nardi S, Reniero F, Dell'Agnola G. Structural characteristics and biological activities of humic substances within the Ah horizon (Calcic-Luvisol). Plant and Soil. 1996;179:65-72.

[35] Quaggiotti S, Ruperti B, Pizzeghello D, Francioso O, Tugnoli V, Nardi S. Effect of low molecular size humic substances on nitrate uptake and expression of genes involved in nitrate transport in maize (*Zea mays* L.). Journal of Experimental Botany. 2004;55:803-813.

[36] Nunes OR, Domiciano GA, Alves WS, Melo ACA, Sousanogueira FC, Canellas LP, Olivares FL, Zingali RB and Soares MR. Evaluation of the effects of humic acids on maize root architecture by label-free proteomics analysis. Scientific Reports. 2019;9:12019. https://doi.org/10.1038/s41598-019-48509-2.

[37] Canellas LP, Olivares FL, Okorokova-Facanha AL, Facanha AR. Humic acids isolated from earthworm compost enhance root elongation, lateral root emergence, and plasmamembrane H-ATPase activity in maize roots. Plant Physiology. 2002;Plant Vol.130, pp. 1951-1957.

[38] Mora V, Baigorri R, Bacaicoa E, Zamarreno A M and Garcia-Mina JM. The humic acid-induced changes in the root concentration of nitric oxide, IAA and ethylene do not explain the changes in root architecture caused by humic acid in cucumber. Environmental and Experimental Botany.2012;76:24-32.

[39] Nardi S, Pizzeghello D, Schiavon M and Ertani A. Plant biostimulants: Physiological responses induced by protein hydrolyzed-based products and humic substances in plant metabolism. Scientia Agricola. 2016;73(1):18-23. http://dx.doi.org/10.1590/0103-9016-2015-0006.

[40] Mora V, Bacaicoa E, Zamarreño AM, Aguirre E, Garnica M, Fuentes M. Action of humic acid on promotion of cucumber shoot growthinvolves nitrate-related changes associated with the root-to-shoot distributionof cytokinins, polyamines and mineral nutrients. Journal of Plant Physiology. 2010;167:633-642. DOI: 10.1016/j.jplph.2009.11.018.

[41] Yildirim E. Foliar and soil fertilization of humic acid affect productivity and quality of tomato. Acta Agriculturae Scandinavica Section B-Soil and Plant Science. 2007;57: 182186.

[42] Nikbakht A, Kafi M, Babalar M, Xia YP, Luo A, and Etemadi N Effect of humic acid on plant growth,nutrient uptake, and postharvest life of gerbera. Journal of Plant Nutrition. 2008;31:2155-2167.

[43] Man-hong Y, Lei Z, Sheng-tao X, McLaughlin NB, Jing-hui L. Effect of

water soluble humic acid applied to potato foliage on plant growth, photosynthesis characteristics and fresh tuber yield under different water deficits. Scientific Reports. 2020;10:7854. https://doi.org/10.1038/s41598-020-63925-5.

[44] Kıran S, Furtana GB, Talhouni M, Ellialtıoğlu SS. Drought stress mitigation with humic acid in two *Cucumis melo* L. genotypes differ in their drought tolerance. Bragantia, Campinas. 2019;78(4):490-497. DOI: https://doi.org/10.1590/1678-4499.20190057.

[45] García AC, Santos LA, Izquierdo FG, Sperandio MVL, Castro RN, Berbara RLL. Vermicompost humic acids as an ecological pathway to protect rice plant against oxidative stress. Ecological Engineering. 2012;47:203-8.

[46] Lotfi R, Kalaji HM, Valizadeh GR, Khalilvand Behrozyar E, Hemati A, Gharavi-kochebagh P, Ghassemi A. Effects of humic acid on photosynthetic efficiency of rapeseed plants growing under different watering conditions. Photosynthetica 2018:56(3):962-970. DOI: 10.1007/s11099-017-0745-9.

[47] Abhari A, Gholinezhad E. Effect of humic acid on grain yield and yield components in chickpea under different irrigation levels. Journal of Plant Physiology and Breeding. 2019; 9(2):19-29.

[48] Canellas LP, Canellas NAO, da S Irineu LES , Olivares FL, Piccolo A. Plant chemical priming by humic acids. Chemical and Biological Technologies in Agriculture. 2020;7:12. https://doi.org/10.1186/s40538-020-00178-4.

[49] Kaya C, Akram NA, Ashraf M, Sonmez O. Exogenous application of humic acid mitigates salinity stress in maize (*Zea mays* L.) plants by improving some key physico-biochemical attributes. Cereal Research Communications. 2018;46(1);67-78. DOI: 10.1556/0806.45.2017.064.

[50] Turkmen O, Dursun A, Turan M, Erdinç C. Calcium and humic acid affect seed germination, growth and nutrient content of tomato (*Lycopersicon esculentum* L.) seedlings under saline soil conditions. Acta Agriculturae Scandinavica, Section B — Soil & Plant Science. 2004;54, 168-174.

[51] Khaleda L, Park HJ, Yun D, Jeon J, Kim M, CHA JY and Kim W. Humic acid confers high-affinity K^+ transporter 1-mediated salinity stress tolerance in *Arabidopsis*. Molecules and Cells. 2017;40(12):966-975

[52] Cha JY, Kang SH, Ali I, Lee SC, Ji MG, Jeong SY, Shin GI, Kim MG, Jeon JR, Kim WY. Humic acid enhances heat stress tolerance via transcriptional activation of Heat-Shock Proteins in *Arabidopsis*. Scientific Reports. 2020;10:15042. https://doi.org/10.1038/s41598-020-71701-8

[53] Weber J. Humic substances and their role in the environment. EC Agriculture. 2020;03-08.

[54] van Rensburg CE. The anti-inflammatory properties of humic substances: A mini review. Phytotherapy Research. 2015;29(6):791-5. doi: 10.1002/ptr.5319.

[55] Carlos Carrasco-Gallardo, Leonardo Guzman, and Ricardo B. Maccioni. Shilajit: A Natural Phytocomplex with Potential Procognitive Activity. International Journal of Alzheimer's Disease. 2012; Article ID 674142, 4 pages doi:10.1155/2012/674142

[56] Klocking R, Helbig B. Medical aspects and applications of humic substances. Biopolymers for medical and pharmaceutical applications. WILEY-VCH Verlag GmbH and C. KGaA. Weinheim; 2005. p. 3-16.

[57] Beer AM, Junginger HE, Lukanov J, Sagorchev P. Evaluation of the permeation of peat substances through human skin in vitro. International Journal of Pharmaceutics. 2003;253 (1-2):169-75. DOI:10.1016/s0378-5173 (02)00706-8.

[58] Ma'or Z, Henis Y, Alon Y, Orlov E, Sørensen KB, Oren A. Antimicrobial properties of dead sea black mineral mud. International Journal of Dermatology. 2006;45(5):504-11. DOI: 10.1111/j.1365-4632.2005.02621.x.

[59] Ghosal S, Singh SK, Kumar Y, Srivastava R, Goel RK, Dey R. Anti-ulcerogenic activity of fulvic acids and 4′-methoxy-6-carbomethoxybiphenyl isolated from shilajit. Phytother Res. 1988;2:187-91

[60] van Rensburg CEJ, Badenhorst BE, Gandy JJ and Snyman JR. Potassium humate reduces inflammation and clinically improves the outcomes of patients with osteoarthritis of the knee. The Open Conference Proceedings Journal. 2010;1:69-74.

[61] Balkrishna, A., Sakat, S.S., Joshi, K. et al. Herbo-mineral formulation 'Ashwashila' attenuates rheumatoid arthritis symptoms in collagen-antibody-induced arthritis (CAIA) mice model. Sci Rep. 2019;9:8025. https://doi.org/10.1038/s41598-019-44485-9

[62] Islam K M, Schumacher S A, Gropp J M. Humic acid substances in animal agriculture. Pakistan J. Nutr. 2005;4:126-134. 10.3923/pjn.2005.126.134

[63] El-Zaiat HM, Morsy AS, El-Wakeel EA, Anwer MM and Sallam SM. Impact of humic acid as an organic additive on ruminal fermentation constituents, blood parameters and milk production in goats and their kids growth rate. Journal of Animal and Feed Sciences. 2018; 27:105-113.

[64] Gerlach H, Gerlach A, Schrödl W, Schottdorf B, Haufe S. Oral application of charcoal and humic acids to dairy cows influences clostridium botulinum blood serum antibody level and glyphosate excretion in urine. Journal of Clinical Toxicology. 2014;4:186. DOI: 10.4172/2161-0495.186.

[65] Sahin A, Iskender H, Kapakin KA Terim, Ainkaynak K, Hayirli A, Gonultas A, Kaynar O The effect of humic acid substances on the thyroid function and structure in lead poisoning. Revista Brasileira de Ciência Avícola. 2016;18(4):649-654. https://doi.org/10.1590/1806-9061-2016-0299.

[66] Jusadi D, Aprilia T, Setiawati M, Suprayudi MA, Ekasari J. Dietary supplementation of fulvic acid for growth improvement and prevention of heavy metal accumulation in Nile tilapia fed with green mussel. The Egyptian Journal of Aquatic Research. 2020;46(3):295-301. https://doi.org/10.1016/j.ejar.2020.04.002.

[67] Hayirli A, Esenbuga N, Macit M, Lacin E and Karaoglu M, Karaca H and Yildiz L. Nutrition Practice to Alleviate the Adverse Effects of Stress on Laying Performance, Metabolic Profile, and Egg Quality in Peak Producing Hens: I. The Humate Supplementation. Asian-Australasian Journal of Animal Sciences. 2005;18(9): Doi:10.5713/ajas.2005.1310.

[68] Cetin E, Guclu B K, Cetin N. Effect of Dietary Humate and Organic Acid Supplementation on Social Stress Induced by High Stocking Density in Laying Hens. Journal of Animal and Veterinary Advances. 2011;10(8): 2402-2407 DOI:10.3923/javaa.2011.2402.2407

[69] Lieke T, Steinberg CEW, Pan B, Perminova IV, Meinelt T, Knopf K, Kloas W. Phenol-rich fulvic acid as a water additive enhances growth, reduces stress, and stimulates the immune system of fish in aquaculture.

Scientific Reports. 2021;11.174. DOI:10.1038/s41598-020-80449-0.

[70] Ozkan A, Sen HM, Sehitoglu I, Alacam H, Guven M, Aras AB, Akman T, Silan C, Cosar M, Karaman HIO. Neuroprotective effect of humic acid on focal cerebral ischemia injury: An experimental study in rats. Inflammation. 2015;38:32-39.

[71] Yang GE, Tae HJ, Lee TK, Park YE, Cho JH, Kim DW, Park JH, Ahn JH, Ryoo S, Kim YM, Shin MC, Cho JH, Lee CH, Hwang IK, Jin H, Won MH, Lee JC. Risperidone treatment after transient ischemia induces hypothermia and provides neuroprotection in the gerbil hippocampus by decreasing oxidative stress. International Journal of Molecular Sciences, 2019;20(18):4621. https://doi.org/10.3390/ijms20184621.

[72] Akbas, A, Silan, C, Gulpinar MT. Renoprotective effect of humic acid on renal ischemia-reperfusion injury: an experimental study in rats. Inflammation. 2015;38:2042-2048. https://doi.org/10.1007/s10753-015-0185-2

[73] Cagin YF, Sahin N, Polat A, Erdogan MA, Atayan Y, Eyol E, Bilgic Y, Seckin Y, Colak C. The acute effect of humic acid on iron accumulation in rats. Biological Trace Element Research. 2016;171:145-155.

[74] Shikalgar TS and Naikwade NS. Evaluation of cardioprotective activity of fulvic acid against isoproterenol induced oxidative damage in rat myocardium. International Research Journal of Pharmacy. 2018;9(1): 71-80.

[75] Gao Y, He J, He Z. Effects of fulvic acid on growth performance and intestinal health of juvenile loach *Paramisgurnus dabryanus* (Sauvage). Fish & Shellfish Immunology. 2017;62:47-56. DOI: 10.1016/j.fsi.2017.01.008.

[76] Aydin SK, Dalgic S, Karaman M, Kirlangic OF, Yildirim H. Effects of fulvic acid on different cancer cell lines. *Proceedings.* 2017;*1:*1031. https://doi.org/10.3390/proceedings1101031.

[77] Ting HC, Yen CC, Chen WK, Chang WH, Chou MC, Lu FJ. Humic acid enhances the cytotoxic effects of arsenic trioxide on human cervical cancer cells. Environ Toxicol Pharmacol. 2010;29(2):117-25. doi: 10.1016/j.etap.2009.11.009.

[78] Pant K, Gupta A, Gupta P, Ashraf A, Yadav A, Venugopal S. Anti-proliferative and anticancer properties of fulvic acid on hepatic cancer cells. J Clin Exp Hepatol 2015;5:S2. DOI: https://doi.org/10.1016/j.jceh.2015.07.005

[79] Aykac A, Becer E, Okcanoğlu TB, Güvenir M, Süer K, Vatansever S. The cytotoxic effects of humic acid on human breast cancer cells. *Proceedings.* 2018;2(25):1565.

[80] Tavassoli A, Monsefi M. The anti-cancer property of mumie as natural product on human cervical cancer cell line (HeLa). Journal of Environmenatl Treatment Techniques. 2020;9(1):196-202. https://doi.org/10.47277/JETT/9(1)202.

[81] Kaevska M, Lorencova A, Videnska P, Sedlar K, Provaznik I, Trckova M. Effect of sodium humate and zinc oxide used in prophylaxis of post-weaning diarrhoea on faecal microbiota composition in weaned piglets. Veterinarni Medicina. 2016;61:328-336.

[82] Visscher C, Hankel J, Nies A. Performance, fermentation characteristics and composition of the microbiome in the digest of piglets kept on a feed with humic acid-rich peat. Frontiers in Veterinary Science. 2019;6:29. DOI:10.3389/fvets.2019.00029.

[83] Arafat RY, Khan SH, Abbas G, Iqbal J. Effect of dietary humic acid via drinking water on the performance and egg quality of commercial layers. American Journal of Biological and Life Sciences. 2015;3:26-30.

[84] Arif M, Alagawany M, Abd El-Hack ME, Saeed M, Arain MA, Elnesr SS. Humic acid as a feed additive in poultry diets: a review. Iranian Journal of Veterinary Research. 2019;20(3):167-172.

[85] Chang Q.F., Bai H.X., Shi B.M., Shan A.S., Wei C.Y., Yu C.Q., Tong B.S. Effects of dietary FA on the growth performance, serum biochemical indices, routine blood parameter and immunity of growing swine. Chin. J. Anim. Nutr. 2013;25:1836-1842.

[86] Mao, Y. Modulation of the Growth Performance, Meat Composition, Oxidative Status, and Immunity of Broilers by Dietary Fulvic Acids. Poult. Sci. 2019;10:4509-4513. https://doi.org/10.3382/ps/pez281

[87] Vucskits AV, Hullár I, Bersényi A, Andrásofszky E, Kulcsár M, Szabó J. Effect of fulvic and humic acids on performance, immune response and thyroid function in rats. Journal of Animal Physiology and Animal Nutrition (Berl). 2010;94(6):721-8. DOI: 10.1111/j.1439-0396.2010.01023.x.

[88] Gomez-Rosales S, de L. Angeles M. Addition of a worm leachate as source of humic substances in the drinking water of broiler chickens. Animal Bioscience 2015;28(2):215-222. DOI: https://doi.org/10.5713/ajas.14.0321.

[89] Zhernov YV, Kremb S, Helfer M, Schindler M, Harir, M Mueller C, Hertkorn N, Avvakumova NP, Konstantinov AI, Brack-Werner R, Schmitt-Kopplindg P and Perminova IV (2017) Supramolecular combinations of humic polyanions as potent microbicides with polymodal anti-HIV-activities. New J. Chem. 2017;41:212-224

[90] Zhernov YV, Konstantinov AI, Zherebker A, Nikolaev E, Orlov A, Savinykh MI, Kornilaeva GV, Karamov EV, Perminova IV. Antiviral activity of natural humic substances and shilajit materials against HIV-1: Relation to structure, Environmental Research. 2021; 193:110312.https://doi.org/10.1016/j.envres.2020.110312.

[91] Kornilaeva GV, Siniavin AE, Schultz A. The differential Anti-HIV effect of a new humic substance-derived preparation in diverse cells of the immune system. Acta Naturae. 2019;11(2):68-76. DOI: 10.32607/20758251-2019-11-2-68-76.

[92] Hafez M, Popov AI, Zelenkov VN, Teplyakova TV and Rashad M. Humic substances as an environmental-friendly organic wastes potentially help as natural anti-virus to inhibit COVID-19. Science Archives. 2020;1(2):53-60. http://dx.doi.org/10.47587/SA.2020.1202.

[93] Winkler J, Ghosh S Therapeutic potential of fulvic acid in chronic inflammatory diseases and diabetes. Journal of Diabetes Research. 2018;Article ID 5391014, 7 pages. https://doi.org/10.1155/2018/5391014

[94] Murbach TS, Glávits R, Endres JR, Clewell AE, Hirka G, Vértesi A, Béres E, Pasics Szakonyiné I. A toxicological evaluation of a fulvic and humic acids preparation. Toxicology Reports. 2020;7:1242-1254.

[95] Dai C, Xiao X, Yuan Y, Sharma G, Tang S. A comprehensive toxicological assessment of fulvic acid. Evidence-based complementary and alternative medicine:eCAM, 2020;8899244. https://doi.org/10.1155/2020/8899244.

[96] Hseu YC, Chen SC, Chen YL, Chen JY, Lee ML, Lu FJ, Wu FY, Lai JS, Yang HL. Humic acid induced genotoxicity in human peripheral blood lymphocytes using comet and sister

chromatid exchange assay. Journal of Hazardous Materials. 2008;153(1-2):784-91. DOI: 10.1016/j.jhazmat.2007.09.024.

[97] Huang TS, Lu FJ, Tsai CW, Chopra IJ. Effect of humic acids on thyroidal function. J Endocrinol Invest. 1994;17(10):787-91. doi: 10.1007/BF03347776.

[98] Cooksey R C, Gaitan E, Lindsay R H, Hill J B and Kelly K. Humic substances, a possible source of environmental goitrogens. Organic Geochemistry. 1985;8(1):77-80. https://doi.org/10.1016/0146-6380(85)90054-3

[99] Laurberg, P., Andersen, S., Pedersen, I.B., Ovesen, L. and Knudsen, N. Humic substances in drinking water and the epidemiology of thyroid disease. BioFactors. 2003;19:145-153. https://doi.org/10.1002/biof.5520190307

Chapter 2

Dry Cowdung Powder - Novel Unearthed Humus: Sustains Water-Food-Energy Nexus

Hemlata K. Bagla

Abstract

Water-Food-Energy (WFE) Nexus regulates biosphere and irrespective of alphabetical or chronological order, it must have synergy for the sustainability of life. Population, Pollution and current Pandemic- COVID has made it vivid to entire scientific community that unless we strengthen this trio, our progressive humanity is sure to collapse. Humic Substances (HS), the originator of the life has promising utilization in almost every sectors of life, reinforcing WFE Nexus. This chapter dedicates to novel unearthed HS, Dry Cowdung Powder (DCP) and its unique contribution in sustaining the triangle of life, WFE. DCP has been employed as humiresin for bioremediation of wastewater containing heavy metals and radionuclides. The known candidate for Biorhexistasy, DCP increases soil fertility, minimizes erosion and acidification. It is also extensively explored as biofuel, green and clean source of energy. 'Nothing of Nature is a Waste' and 'Waste is a Commodity' aptly describes DCP, the sustainer of WFE Nexus.

Keywords: Sustainability, Dry Cowdung Powder, Unearthed Humus, Humiresin, Biorhexistasy, Biofuel

1. Introduction

The biosphere or Earth's layer of Living Matter harbors living organisms that interact with each other as well as with the soil, water, and air. The necessity of the entire living biota is based on Water-Food-Energy (WFE) Nexus, exceptionally governed by our Mother Nature. In the past few centuries, rapid industrialization has overexploited pristine ecosystems and led to a steady decline in the nutrient quality of ecosystems, flora, and fauna. The plume of population and pollution has created a 'water-food hungry scenario' but the 21st century socio-political race has created 'energy hungry' attitude too. The high rate of resource consumption has posed an unintended and negative impact not only on the entire ecosystem but also on global geopolitical harmony. Almost every nation has gobbled up their own abundance of natural resources due to utilitarian attitude and drive of being energy independent. This has compromised the holistic functioning of the environment and given rise to deprived soils and declining fauna.

Present resource crisis is so all-prevailing that, both developed as well as developing countries have come to the common conclusion of Sustainability through Socio-Scientific Integrity for the smooth survival of todays and tomorrows.

Nonetheless, current pandemic Covid-19 has sparked SOS to refurbish the global policies of governance and accelerate scientific eclecticism towards sustaining the WFE Nexus by participating in Sustainable Ecosystem Management (SEM). The philosophy of SEM is pegged on adaptive management and holistic consideration of ecological resources. It is environmentally sensitive, ecosystem, and eco-regional based [1]. It emphasizes on balancing long-term sustainability of ecosystems with human needs and moves beyond compartmentalized focus on some species and regions, to holistic management of natural resources amidst hindrances of rising resource of demand, population increase, and climate variability [2, 3].

The aim of this study is to evaluate, equate and exercise the application of novel humiresin- unearthed, ecology-economy friendly, zero-waste prototype Dry Cowdung Powder (DCP) towards sustaining WFE Nexus and ecosystem.

2. WFE Nexus: humic substances

2.1 Sustainability through humic substance fecundity

Sustainability is not the futuristic projection of survival but is the continuous mechanism of proliferation just like the evolution of life. The evolutionary journey from the prebiotic world to today's most complex homochiral D-sugars, L-amino acids organisms; we sapiens, are the product of spatial characteristics and elemental selectivity of Humic Substances (HS), the originator of life! [4] The first organic machinery in proplanetary disk, HS is approximately four billion years old and is the bridge which connects no-life to life [5]. The Miller's experiments gave the first proof of HS as the placenta of evolution and this theory has been supported by many researchers [6, 7]. Owing to its universal properties, HS are popularly known as 'the black gold of the earth' by farmers and 'the black box' by soil analysts [8].

Soil is an ecosystem as a whole and acts as a modifier of planet Earth's atmosphere. The organic matter present in it originates mainly from plant metabolites with the onset of senescence and humification. Humic Substances are defined as a heterogeneous mixture of morphologically changed degradation products of biomolecules which on humification give rise to a family of amorphous, polyelectrolytic, polydisperse and colloidal compounds [9]. HS are universally recognized as chemically, biologically, and physically active due to the typical composition, macromolecular structure, polyfuntionality, surface properties, and presence of multiple reactive sites, variable size-shapes, and intrinsic porosity [10]. Operationally, HSs are usually extracted at high pH and classified into three categories based on their solubility: acid-soluble Fulvic acid (FA), acid-insoluble Humic acid (HA), and base-insoluble Humins with the average properties and a large assembly of components with diverse structure and molecular weight [11].

Humic substances exhibit exceptional characteristics for soil conditioning. HS have shown to contain auxin-like activities [12] and hence they are successfully used in agronomy, environmental, industrial, medicinal, and pharmaceutical fields [13, 14]. Besides providing nutrients and aeration to the soil, they interact and bind with toxic heavy metals, radionuclides, pesticides, industrial dyes, and other xenobiotics that may be present as pollutants in the ecosystem, thus acting as natural sieves. HS mimics geochemical barrier and hence potentially can be used as a filling material for barrier walls to prevent transport and bioavailability of heavy metals in soil [15]. HS can significantly reduce the acute toxicity of metals and bioavailability of metals. It plays a crucial role in the inactivation of pesticides, heavy metals, and other polluting agents. The various studies on HS, have confirmed a great potential for an application in aerobic and anaerobic wastewater treatment as well as in bioremediation [16]. Study of HS has proven to be a robust approach in resolving environmental issues and developing meaningful paths for modification steps [17]. HS provides insights into pollutant mobility, behavior, and fate, and allows better understanding of natural mitigation phenomena and potential impact on the environment [18].

Our global scientific community has univocal conscience on how amazing and myriad utilization of HS have been fructified in almost every sectors of life since the inception of life. Being ubiquitous in both terrestrial and aquatic niche, HS is considered to play a crucial role in sustaining the ecosystem.

2.2 Managing nexus through humus and waste

Entire ecosystem is governed by many layers and sectors, inter and intra connected, independent yet mutually operating native ecosystems. The concept of WFE Nexus is very much like the vertices of an equilateral tripod, regulating the survival of life. Irrespective of its alphabetical or chronological order, there is a need to govern and manage inter and intra disciplinary synergy to ensure fundamental and survival rights of clean air, food, and water to our future generations. Governing the Nexus is probably one of the grand challenges of current time, as quoted by Jeremy Allouche et.al, explicitly explaining the reason for complexity in the management of so called 'nirvana concept' of nexus [19]. Author's justification on 'Nexus or Nexuses?' indicates that holistic, comprehensive, and integrated implementation of the scientific aptitude is required to achieve the goal of global sustainability. Nexuses, is a conceptual approach that does not hold rigidly to a single sector or a paradigm or set of assumptions. Instead, it appeals upon multiple scientific theories or principles to gain complementary insights according to geography and climate of the sector. It does not appreciate the 'silo' concept or 'two at once' methodology to address the sustainability issues, but focuses on simultaneous operating approach.

A review by A. Hamidov and K. Helming [20] rationalizes that to improve cross-sectoral coordination in support of sustainable development, WFE nexus has become an important concept in natural resource management. On the same note, Zhang et al. [21] suggests, shifting the focus to WFE nexus is valuable but downplaying their interrelations can have adverse consequences on the entire biota. Hence, there is a need for the balanced approach with 'no-regrets' solution as studied by Konadu et al. [22].

Resources produce waste and it is undoubtedly inseparable from any sector of life, omnipresent both inside and outside of our existence. Globally, hulk of waste is directly or indirectly connected to production of humus and hence HS plays a very important role in Waste Management (WM). WM [23] is the collection, transport, processing, monitoring, and recycling or disposal of the waste material. It is based on waste hierarchy [24] of reduce, reuse, and recycle coined with re-think. The effective implementation and maintenance of stipulated laws requires legions of regulators. It offers an efficient means of managing waste that fits ideally with the needs of large population centers, while providing a source of alternative and renewable energy [25]. Waste-to-Energy (WTE) is a philosophy of energy production using the waste and its role in enhancing all interconnected nexuses, is gaining high importance in every nation [26, 27]. 'Nothing of Nature is a Waste' and 'Waste is Green Commodity' aptly describes DCP and our research on its application proves its pivotal role in sustaining the Nexus.

2.3 Unearthed humus DCP- zero waste prototype

The pedosphere or soil body is rich in organic matter, soil fauna, minerals, water, gases, that together support life. It comprises of weathered rocks from the lithosphere, trapped air pockets from the atmosphere, moisture from the hydrosphere, along with decomposing matter from the biosphere, and is thus known as an interface between these subsystems. Besides purifying water in terrestrial systems, soils process decaying matter from the biosphere and recycle nutrients back into the food chain. The pedosphere thus interconnects with other spheres and regulates life processes.

Humus, the organic matter in soil, permits better aeration, enhances the absorption, releases nutrients, and makes the soil less susceptible to leaching and erosion, thus the agent of soil's vitality. Soils contain more organic carbon than the atmosphere and vegetation combined; therefore, mineralization of soil organic matter and release of carbon versus humification processes, could greatly affect carbon capture and stabilization, and consequently the global climate change. Carbon sequestration and storage in soils provide an important means of reducing greenhouse gases in the atmosphere to mitigate predicted climate changes [28].

The aforementioned qualities of HS make it a very powerful tool not only in sustenance of WFE Nexus but its essential role in mitigation of climate change. This draws our attention towards requirement of highly appropriate method for the Selection of HS products as there are enormous variety of HS available globally. The quality and composition of any mined or extracted Humates or HS depends on many different factors as HS are essentially time and climate dependent.

Under ideal conditions, the turnover time of organic carbon added in form of HS each year from plant and animal residues into the soil, averages approximately 30 years [29]. Due to this time constraint in production of fresh humus, we face many challenges ovulated due to multiple sites and over digging of soils. Also, soil humus development is highly affected by seasonal change, litter type and time, hot or cold climate and freeze–thaw cycle [30]. Studies by Dou et al. have explained that freeze–thaw cycles can even destroy the structure of newly formed humus [31]. Nonetheless, a good HS material can be destroyed by improper mining or processing. Hence the mining and the extraction of HS must be meticulously processed as any error or shift in parameters may change their physicochemical properties.

Thus, it is fundamentally important to extract them with minimal change into its original structure [32]. The physical process of removing the ores or materials from the ground is disruptive to the environment and alarming questions regarding the statutory norms for safety are overwhelming. Also, the pre and post process for concentration of raw material and final product is time consuming and makes an essential call for the higher degree of purification. This leads to the generation of many undesirable chemical steps, adding to cost of actual extraction process as well as stress on environmental safety.

To solve this complicated situation, we bring the concept of Unearthed Humus, DCP – clean, economical, and essentially inert to any climatic condition, due to its generation in to the biotic environment of cow's rumen. It is with least contaminants as the process of Humification takes place during the period of 1 month of Sun drying, hence conditioning of any type is not required thanks to its biological matrix. These qualities of DCP projects the Zero Waste concept of Waste Management, boosting Green Chemistry principles. Zero Waste Management is the holistic concept of waste management which recognizes waste as a resource produced during the interim phase of the process of resource consumption [33]. The prime objective of this protocol is based on the sustainable avoidance and management of waste and resources. This makes us very positive to conclude that Unearthed DCP is the Zero Waste Prototype in comparison with all the HSs used at present.

2.4 Green solution to soil pollution: DCP- agrarian booster

Soil is a heavenly resource, a living, breathing and ever-changing dynamic ecosystem. The theory of Biorhexistasy, describes two crucial climatic phases of soil i.e., Biostasy, period of soil formation and Rhexistasy, periods of soil erosion [34]. The theory emphasizes on the role of climatic conditions that initiate pedogenesis and subsequent soil retrogression. For example, moist climates promote the chemical deterioration of rocks, leading to formation of soils and improvement in vegetation, which reflects the period of biostasy. On the other hand, dry climates and winds bring about displacement of soils and mechanical disintegration of exposed rocks, leading to the unfavorable period of rhexistasy or soil erosion and degradation. The two stages occur in cyclic form and control soil quality and vegetation. Extended periods of rhexistasy lead to ever-increasing soil degradation.

Thus, due to soil erosion, salination, and extreme acidification, large portions of arable land lose their fertility. The scale of this agrarian distress can be improved by the addition of HA, which promotes root growth and induces salt tolerance in plants, thus reversing the adverse effects of salt stress [35].

For economizing the addition of HA into soils, DCP can be used since it is freely available and inexpensive. Addition of DCP initiates the biotransformation of saline soils and enhances their humus content. Due to the rich microbial biota and chemically inert properties, DCP has been utilized as a soil conditioner and fertilizer since ancient times. Thus, we reiterate that DCP is soil's Green ambrosia and one of the best contenders for soil replenishing and rejuvenation.

2.5 Waste to green energy: DCP- biofuel

As any other non-conventional and renewable green source of energy, DCP has been explored as biofuel and proves to be pseudo shale, boosting renewable energy sector. Feng et al. have developed a high-performance anode from carbonized cow dung for bio-electrochemical systems [36]. Kumar et al. found that in the generation of electricity, natural cow dung was found more suitable as it generated 150.9 mV Open Circuit Voltage (OCV) [37]. There is a review by Gupta et al., focusing on recent findings on cow dung harnessed for biofuel production and management of environmental pollutants and its usage in different areas such as medicine, agriculture, and industry [38].

2.6 Wastewater remediation: DCP - Humiresin

The research work of Bagla et al., has examined and justified an innovative and concrete role of DCP in the bioremediation of toxic metal pollutants and hazardous radionuclides from the pool of various effluents using Radiotracer technique. Notable heavy metal pollutants like trivalent and hexavalent chromium, cadmium [39], mercury [40], silver [41], zinc [42] have been eliminated from aqueous systems using DCP as a biosorbent. Similarly, radionuclides commonly found in radioactive waste - ^{60}Co, ^{90}Sr and ^{137}Cs [43, 44], have been removed from simulated systems of reactor and reprocessing waste.

Being freely and easily available DCP has an edge over processed natural adsorbent considering their cost, time, and energy efficiency. Also, DCP does not impart any foul odor to treated water, nor does it lead to increase in biomass, unlike other natural adsorbents. DCP used for heavy metal adsorption can be subjected to standard desorption and re-adsorption cycles and finally the spent DCP can be employed as a landfill material in deteriorated soils. Because of its strong chelating properties, any residual metallic ions are well-bonded to DCP and do not leach out under normal environmental conditions.

This study concludes that DCP is affordable and adaptable due to its Combo nature and its acceptability lies in its working mechanism based on HSAB concept of sorption [45]. DCP has a great potential in the field of water decontamination, industrial water treatment and in abatement of water pollution.

3. Dry cowdung powder - DCP: The best of waste

DCP is naturally available bioorganic, complex, polymorphic humified fecal matter of cow and is enriched with minerals, carbohydrates, fats, proteins, bile pigments, aliphatic - aromatic species such as HA, FA, Ulmic acid and Humus etc. Many functional groups such as carboxyl, phenols, quinols, amide, enhance its biosorption properties.

The total characterization of DCP has been carried out for its physical, chemical as well as microbiological properties. DCP was provided by Keshav Shrushti, Research Centre on Cow products (Thane, India). Fresh cow dung was collected by efficient workers and due safety measures were taken to avoid any toxic and heavy metal contamination during collection. Cow dung is basically the feed residues digested by symbiotic bacteria residing within the animal rumen. The net effect of digestion in the rumen is the conversion of dietary materials to a mixture of fatty acids (mainly acetic, butyric and propionic acids), gases (primarily CO_2 and CH_4 which are voided by eructation) and microbial biomass [46]. The innate existence of different microbes, beetles and other dung related arthropods bring about humification of the organic matter present in cow dung [47].

There are different pigments and lipids in cow dung which are related to its color and typical odor. The bile pigment biliverdin is mainly present in cow dung (herbivore) giving it its green color. Also, bile salts give dung its emulsifying properties by which it confers hydrophilic coat to the droplets, otherwise of its hydrophobic nature [48]. It is also flourished with number of microorganisms as well as some classes of Arthropods. To ascertain passive biosorption by dead microbes, it is necessary to have an overall account of microbiological consortium of fresh dung.

Cow dung consists of approximately 60 species of bacteria, including species from the following genera - Bacillus, Actinomycetes, Corynebacterium, Pseudomonas, Cellulomonas, Flavobacterium, Lactobacillus, Serratia, and Alcaligens [49, 50]. It also includes *Escherichia coli* and Staphyloccocus aureus along with roughly 100 species of protozoa and yeasts including Saccharomyces and Candida spp. Cow dung also contains certain fungi like Trichoderma and Aspergillus spp. Due to the profusion of diverse micro flora, it has considerable potential for biodegradation and biotransformation [51].

The presence of petroleum utilizing microbes is indicative of high percentage of Hydrocarbon in the environment and cow dung too has a great abundance of this microbiota [52]. The presence of these microbes is also dependent on the geographical and environmental milieu. This microbial consortium enables cow dung with considerable potentials for biodegradation and biotransformation of oil - petroleum products and other pollutants as well as it further contributes to plant production and in many biogeochemical processes [53].

In order to select any new material for a process, the material should fulfill the theory of 3A's, which stand for affordability, acceptability, and adaptability. DCP is affordable due to its free availability and its supply is not hampered by climatic conditions. Most of every HS available requires pretreatments or conditioning for optimal results, unlike DCP, hence it is more time, energy and cost efficient and hence has an edge over other HS. DCP is acceptable owing to its biological matrix, full of microbe's nutrients and least contaminated by any pollutant or contaminants, thanks to its production in perfectly inert or biological niche of rumen. Its adaptability is learned by its combo nature and heterogeneous functionality due to presence of numerous organic and inorganic ligands. These descriptions with its present global application emboss that DCP is the ultimate *Best of Waste*.

4. Physical and chemical characterization of DCP

4.1 Physical characterization

4.1.1 Proximate analysis

The total characterizations of DCP for its physical and chemical properties have been designed. The physical characterization has been carried out by proximate analysis **Table 1** as per the standard procedure given by American Public Health Association (APHA). The elemental, structural, morphological analysis, and thermal stability of DCP has been conducted at Sophisticated Analytical Instrument Facility- Indian Institute of Technology (SAIF- IIT), Mumbai. Physical parameters such as moisture content, ash content, mesh size, fiber content, etc. have been evaluated. Biochemical analysis of DCP for its amino acid, carbohydrate and other contents has been carried out by Radial Chromatography.

4.1.2 The elemental composition by XRF and CHNSO Analyzer

The DCP has been characterized using XRF technology for its quantitative as well as qualitative elemental composition as shown in **Table 2.** For the complete elemental composition, complimentary to XRF technique, C, H, N, S, (O), has also been obtained, **Table 3** shows the same.

Sr. No.	Physio-chemical parameters	Value	Sr. No.	Physio-chemical parameters	Value
1.	Color and Odor	Brownish, Humus smell	9.	Crude fiber content	13.25%
2.	Texture/ Appearance	Powdered	10.	Total Nitrogen content	7.10%
3.	Mesh size	100	11.	Thermal stability	Stable till 150°C
4.	Solubility	Sparingly soluble in water except soluble sugars	12.	Acid-Alkali resistance	Quite high
5.	Moisture content	10–11% & 8.93%	13.	pH	7.14
6.	Ash content	25.07%	14.	Conductance (mmhos/cm)	2.92
7.	Potash content (kg/ ha)	525.00	15.	% Organic carbon	0.60
8.	Total Carbohydrate content	37.13%	16.	Humus content (kg/ha)	35.84

Table 1.
Proximate analysis of DCP.

Element	%
Na	0.946
Mg	2.853
Al	1.684
Si	22.691
P	3.883
K	3.343
Ca	2.360
Ti	0.329
Mn	0.115
Fe	2.419
Cl	1.56
Cr	0.014

Table 2.
Elemental composition of dry cow dung by XRF.

Element %	Component
3.104	Nitrogen
37.367	Carbon
5.142	Hydrogen
3.432	Sulfur
29.654	Oxygen

Table 3.
C,H,N,S,O analysis of DCP.

4.1.3 Scanning electron microscopy (SEM)

The SEM patterns **Figure 1(a-i)**, shows that DCP has some fibrous structure with some holes and small openings on the surface. Also, the surface of DCP is heterogeneous, rough, porous and with dentations. It shows the presence of cell debris of some prokaryotic cells. The holes and openings on DCP increase the contact area and facilitate the pore diffusion during adsorption. SEM study of DCP also affirms the theory of metal ion diffusion into porous biomass. According to this theory, metal ions can either be present on the surface of the biomass or can permeate into the expanded pores, which make desorption of metal ions difficult [54].

4.1.4 Thermal gravimetric analysis (TGA) and differential thermal analysis (DTA)

For the verification of DCP to be thermally stable, *Thermal Gravimetric Analysis* (TGA) as well as *Differential Thermal Analysis* (DTA) have been carried out. **Figure 2** for TGA reveals that, the moisture content of DCP is 8.928% and it is quite stable till 230°C. After 350°C, the weight loss of DCP is about 50%. This can be attributed to simple process such as drying, or from more complex chemical reactions that liberate gasses, such as structural water release, structural decomposition, carbonate decomposition, sulfur oxidation, and fluoride oxidation. Thus, it can be concluded that DCP is thermally stable.

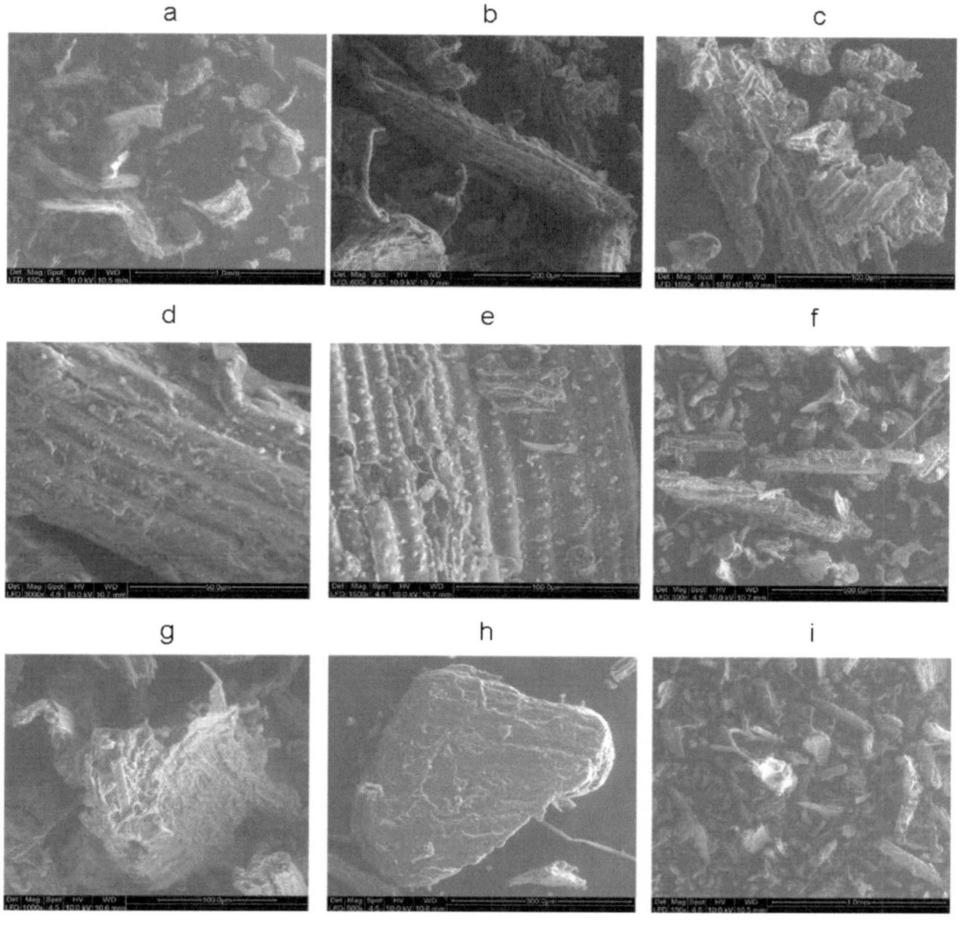

Figure 1.
SEM of DCP.

Figure 3 of DTA shows two endothermic peaks near the region of 48-50°C, which is a small peak and second at 230–270°C is a broad peak. It also describes two exothermic peaks, one smaller peak at 170–175°C and a broader peak from 300°C onwards. This study supports our observation that the biosorption of metal ions on DCP decreases on increasing the temperature. In case of DTA, the mean progression of the combustion profile and the limits of different temperature ranges delineating the phases of thermal decomposition of DCP using 2.311 mg, within a temperature range of 25–420°C with the heating rate of 10°C min^{-1} as can be seen in Graph 2. There are two endothermic and one exothermic peak on curve, corresponding to respective decomposition steps. The first step is in the temperature range between 39.09°C & 60°C and might correspond to the evaporation of water incorporated in or adsorbed onto DCP, being accompanied by a little loss in weight and is signified by first endothermic peak. A small exothermic peak is observed around at 180–185°C, it may be due to some crystallization process involved during heat transfer process [55].

The second endothermic peak is observed in the temperature range from around 214.88–265°C and is accompanied by some weight loss which may be due to loss of polar functional groups. The phenolic OH groups are eliminated between 250°C [56]. Around 280°C the decline in weight is seen which may be caused by decarboxylation and unsaturation [57]. The decomposition of carboxylic, phenolic,

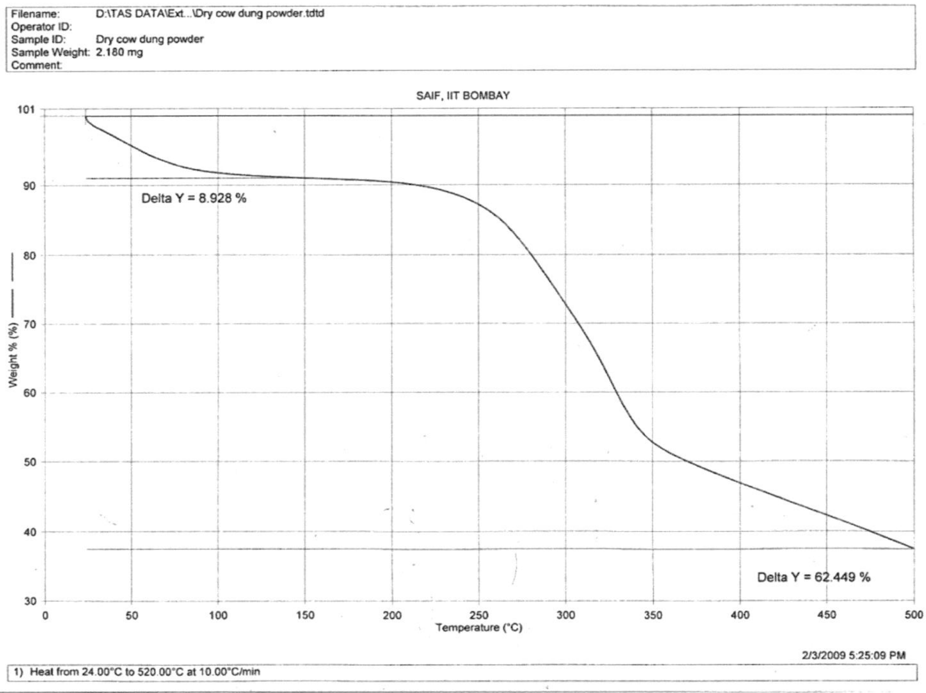

Figure 2.
Thermogravimetric analysis spectra of DCP.

carbonyl and alcoholic groups at higher temperatures have often been attributed to the thermal breakdown of aromatics. In the analytical conditions of TGA and DTA analysis (with a heating rate of 10°C min^{-1}) aromatic structures can be formed from cyclic structures [58].

4.1.5 Electron spin resonance spectroscopy

The abundance and presence of unpaired electron and free radicals in the DCP has been assessed by ESR spectroscopy. The spectral information from **Figure 4** is not very vivid but explains that DCP contains some free radical biotic groups. The source of free radicals can be organic radicals of semi-quinine nature conjugated with extended aromatic systems or paramagnetic metal ions such as Fe, Mn and V [59]. Also, ESR data of IHSS samples do show free radical contents with g-values approximately around 2.000 and we have also obtained the g-value for DCP, around 2.000 in agreement with standard samples.

4.1.6 Fourier-transform infrared spectroscopy (FTIR)

As explained earlier the detailed information of various adsorptive functional groups of biosorbent is of great importance and DCP has been assayed for all the possible functionality present on it. There are various organic groups present on DCP such as carbohydrate, protein, lignin, cellulose etc. In case of carbohydrate, only carboxyl and sulphonate contribute majorly in the formation of metallic ligands. Similarly, in protein moiety the group such as carboxyl, sulphonate,

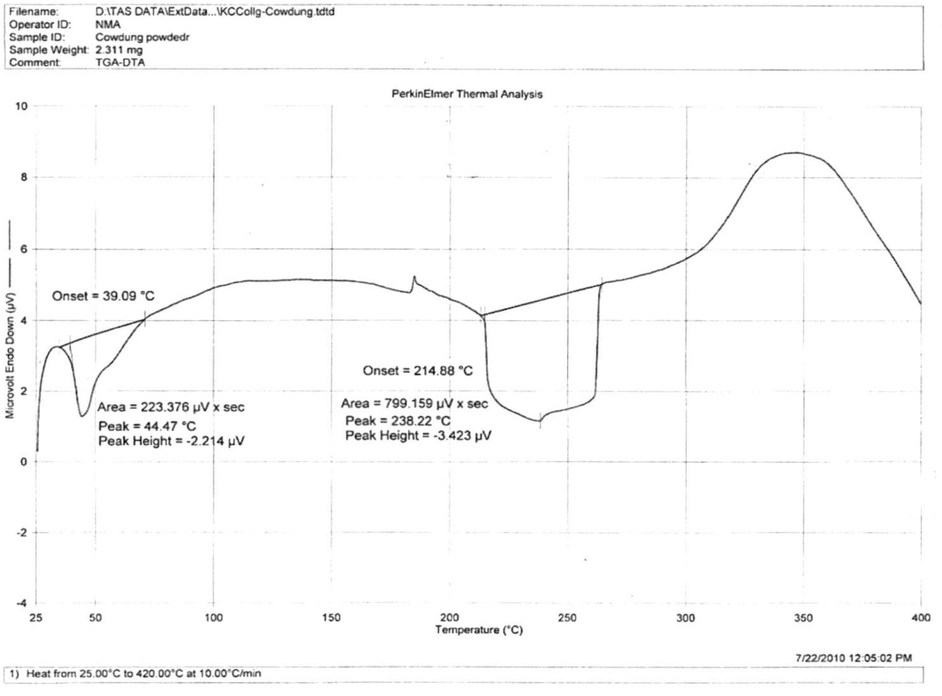

Figure 3.
Differential thermogravimetric analysis spectra of DCP.

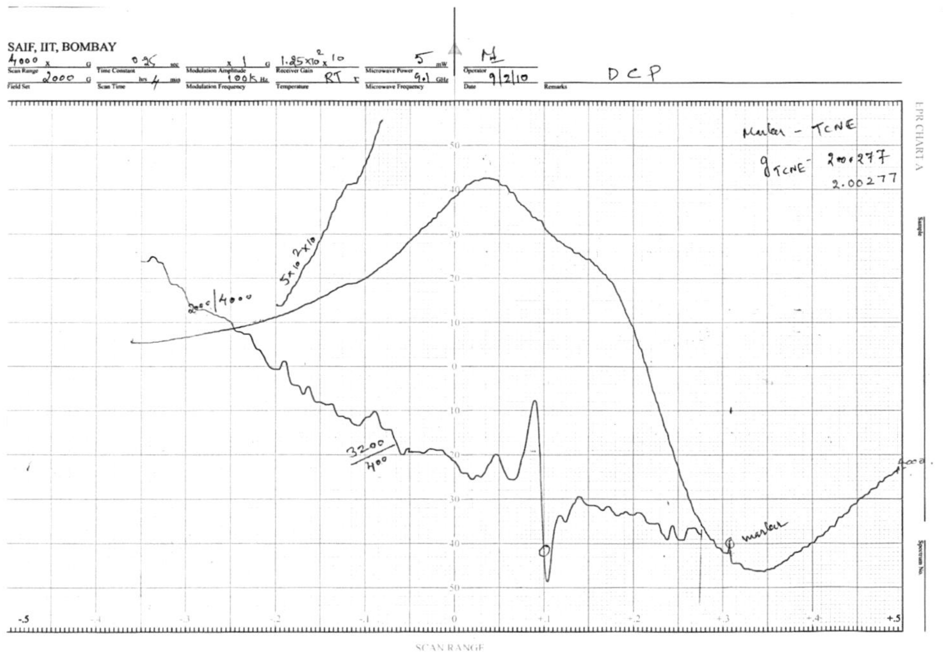

Figure 4.
ESR spectra of DCP.

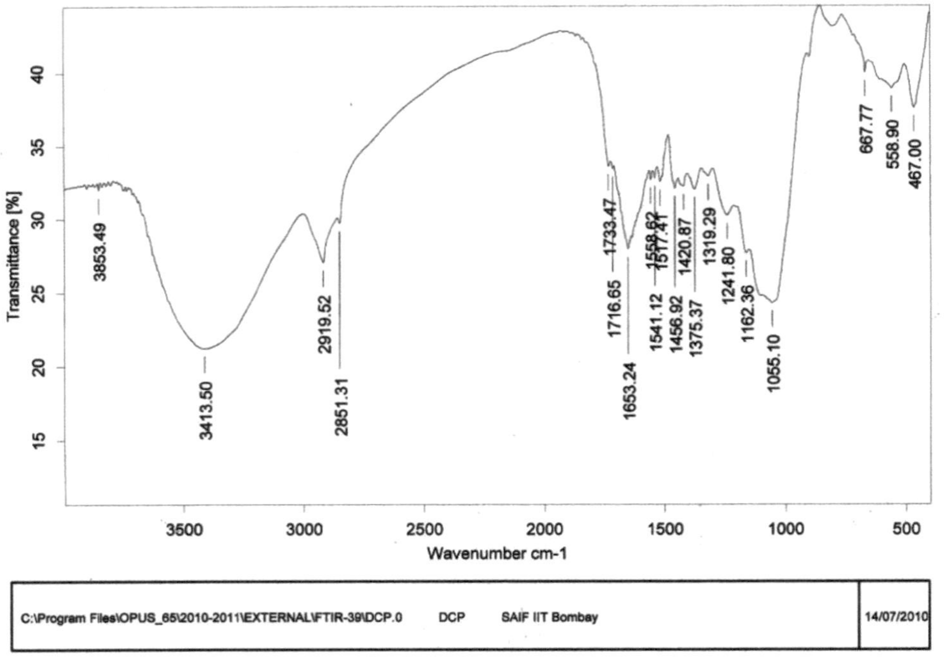

Figure 5.
FTIR spectra of DCP.

sulfhydryl, hydroxyl, phosphonate, thioester, secondary amine, imines participates in ligand formation [60] Lignin derivative contain an abundance of oxygen containing functional group such as phenolic, alcoholic and enolic structure which forms lignin- metal complexes. **Figure 5** and **Table 4** explains the same in brief. Also, the FTIR analysis of DCP after and before the metal ion adsorption has been carried out to confirm the biosorption process with the observed shifts in wavelength of functional group involved in biosorption.

4.2 Chemical assay

4.2.1 Extraction of humic acid from DCP

Literature survey confirms that till today HA has been extracted from the different Abiotic origin worldwide, but a successful pioneering research of extraction from a Biotic Animal origin DCP, has been carried out by authors, employing Green Chemistry Principles as described earlier. We have obtained 9–10% extraction of HA from DCP.

The following technique was devised for extracting HA from dry cow dung powder.

a. For the neutralization of dry cow dung powder, different series of alkali with concentrations ranging from 0.1 to 1.0 M of NaOH, KOH, Na_2CO_3, and $NaHCO_3$ were studied. With 0.1 M NaOH, the best results were obtained.

b. Acids such as HCl, HNO_3, and H_2SO_4 in the molarity range of 0.1–1.0 M were tested for the re-acidification process. With around 0.1 M HCl, the best results were obtained.

c. 15 mL of 0.1 M HCl is desirable for 100 mL of neutralized supernatant.

Functional groups	**Compounds**	**DCP Signal (cm^{-1})**
N-H	Amines	3413.50
-COOH	Carboxylic Acid O-H Stretch	2851.31 2919.52
C-H	Alkane	2851.31 2919.52
C=O & (RCOOR)	Esters & Lactones 6-membered and 5 - membered lactone	1733.47
C=O	Carboxylic acid, (C=O stretch)	1716.65 1733.47
C=OR	6-membered cyclic ketone	1716.65
C=C	Alkene	1653.24
C=C (Stretching Vibrations)	Nonconjugated	1653.24
	Conjugated	1558.62 1541.12
C-H (Bending Vibrations)	CH_2	1456.95 1420.87
	CH_3	1375.37 1319.29
C-Br or Inorganic Impurities	Alkyl halide stretch clay, minerals	558.90 667.77
Si-CH_3	Silicon functions	1241.80
C-O	Saturated secondary or cyclic tertiary amine	1055.10 1162.36
C-Cl	Chlorine	667.77
-Na	metal group	467.00

Table 4.
FTIR data of DCP.

The decantation procedure was used to remove hay and other light non-humic particles by mixing DCP (100mesh) with purified water and stirring for 20 minutes at room temperature. After filtering non-humic materials from cow dung, the resulting slurry was stirred with 0.1 M NaOH for 24 hours. After that, the whole mixture was allowed to settle before being purified. The soluble HA, FA, and other biological matter are found in the filtrate, while the residual contains Humins, Ulmic acid, and insoluble bio-organic matter. Since HA is insoluble at low pH, the filtrate was acidified by slowly applying 0.1 M HCl at room temperature (pH 1) with continuous stirring to mitigate the heat of neutralization and centrifuged for HA precipitation. Fulvic acid and other organic acids are found in centrifugate. To extract HA, the HA was carefully washed with double distilled water until it passed a chloride inspection and dried in an oven at 383 K. We were able to remove 9–10% of the content. FTIR and Raman spectroscopy were used to identify and characterize HA, which was then compared to standard HA from IHSS [61]. At Indian Institute of Technology, Mumbai, we obtained FTIR spectra on a Nicolet Instrument Corporation-USA model-MAGNA 500 with a specification scale of 4000 cm-1 to 50 cm-1. Raman spectra were obtained using a RENISHAW Laser Raman Spectrometer with 325, 514.5, and 785 nm laser excitation and a CCD detector with confocal microscope at the Gemological Institute of India in Mumbai.

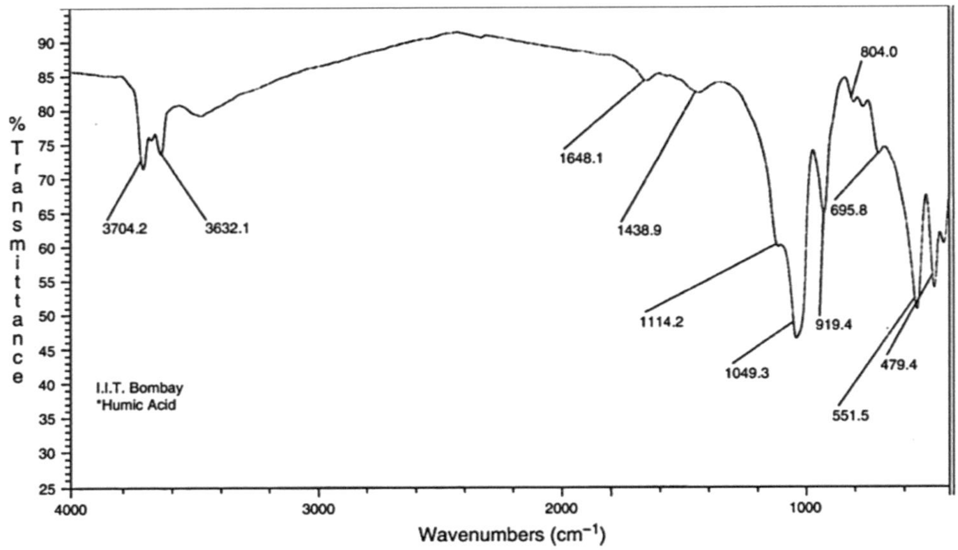

Figure 6.
FTIR spectra of standard HA.

4.2.2 Results and discussion

Two FTIR Spectrum were compared - Standard HA (**Figure 6**) from International Humic Substances Society (IHSS) and HA extracted from DCP (**Figure 7**). **Table 5** elucidates the absorption peaks and their corresponding bonds/ functional groups. Certain shifts in absorption bands may be observed due to inter and intramolecular hydrogen bonding, varying degrees of conjugation, steric hinderances, physical factors and matrix effect. This is mainly because the extracted HA is from a biotic source (cow) and the standard HA has an abiotic origin.

At 3421.0 and 3422.8 cm^{-1}, we observed the band for N-H stretch of primary amine for the standard HA and extracted HA respectively. Whereas the N-H bend

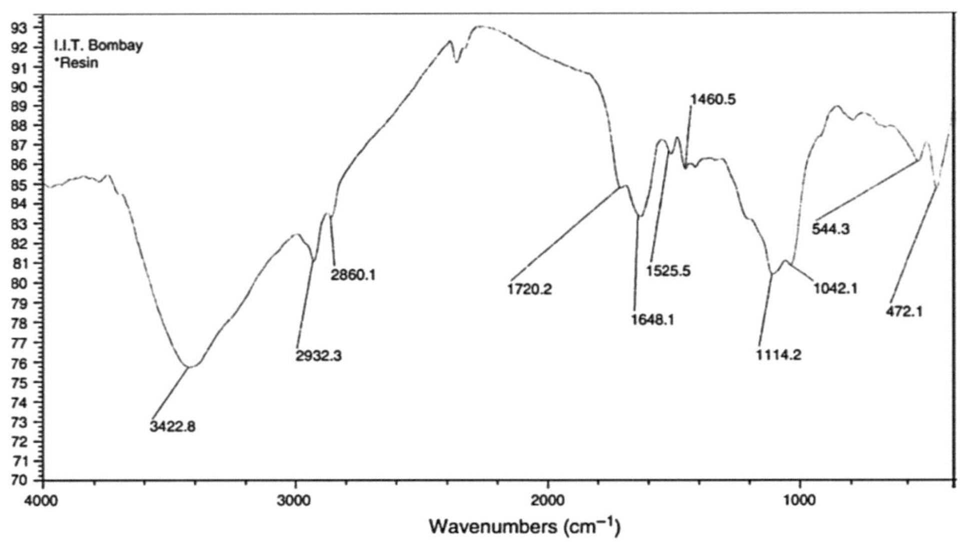

Figure 7.
FTIR spectra of extracted HA.

No.	Functional group	Compounds	Std HA IHSS (cm^{-1})	Ext. HA DCP (cm^{-1})
1.	N-H stretch	Aromatic Monomeric Alcohols, Phenols	3421.0	3422.8
2.	O-H	Carboxylic acid	—	2932.3 2860.1
3.	C=O	Carboxylic acid, Aldehydes, Esters, Ketones	—	1720.2
4.	N-H bend	Primary Amines	1648.1	1648.1
5.	C-N	Aryl or alkyl substituted tertiary amines	1114.2	1114.2
6.	C-O	Primary alcohols	1049.3	1042.1
7.	C-Cl	Chloroalkanes	551.5	544.3

Table 5.
FTIR data of HA.

was seen at 1648.1 cm^{-1}, in both the spectra. Both spectra showed identical bands at 1114.2 cm^{-1} for aryl or alkyl substituted tertiary amines. The band for primary alcohols (C-O) was observed at 1049.3 and 1042.1 cm^{-1} for the standard HA and extracted HA respectively. Similarly, the band for chloroalkanes was observed at 551.5 and 544.3 cm^{-1}, in the two spectra respectively. In the extracted HA, a band is seen at 1720.2 cm^{-1} which is very important as it describes the presence of lactones, which is also influenced by conjugation and ring size and standard HA does not show any band in this region. Thus, we confirm the matrix of extracted HA is from bovine species i.e., fecal residue of cow.

The absorption patterns are depicted in the Raman spectra of standard HA in **Figure 8** and extracted HA in **Figure 9**. **Table 6** lists the absorption bands and their descriptions. Absorption of (C=C) aromatic ring chain heavy vibration was visible at 1580 & 1600 cm^{-1}; ($C-NO_2$) asymmetric mild vibration at 1530–1590 cm^{-1}; (N=N) aliphatic mild vibration at 1550–1580 cm^{-1}; and (X-Metal-O) heavy vibration at 150–450 cm^{-1}. The presence of heavy metals such as iodine, selenium, and silicon is shown by a sharp small absorption band at 150–450 cm^{-1} in the spectra of standard HA. A spectrum of extracted HA is devoid of absorption band at that

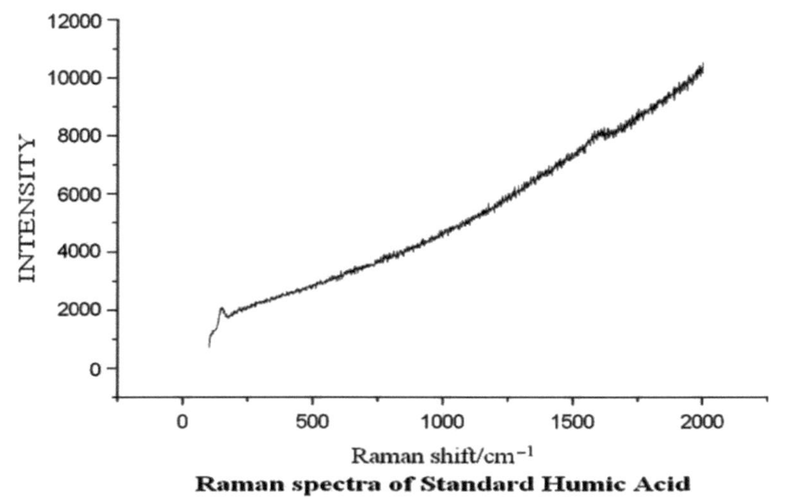

Figure 8.
Raman spectra of standard HA.

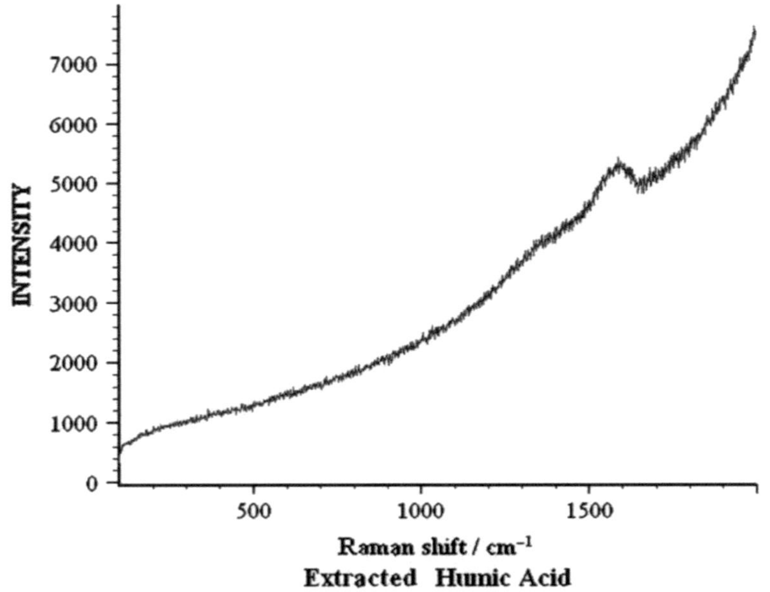

Figure 9.
Raman spectra of extracted HA.

No.	Functional group	Raman shift (cm^{-1})	Band type
1.	C-NO_2	1530–1590	Asymmetrical medium
2.	Aliphatic N=N	1550–1580	Medium
3.	C=N	1610–1680	Medium
4.	C=C	1500–1900	Strong
5.	C=O	1680–1820	Medium
6.	X-Metal-O	150–450	Medium

Table 6.
Raman spectral data of HA.

region which indicates the absence of any contamination of metal ion. This concludes that DCP does not induce any undesirable matrix effect.

4.3 SWOC analysis

SWOC analysis is a strategic planning technique for examining the Strengths, Weaknesses, Opportunities, and Challenges that any project may face. It entails defining the project's goal and determining the internal and external variables that will help or hinder the goal's achievement. Any research project's aim is to move from the lab to the real world. The SWOC analysis of DCP and its application as biosorbent is thoroughly scrutinized to know the feasibility of scaling up in industrial effluent and water treatment process.

4.3.1 Strengths

- DCP is a non-conventional, biodegradable, renewable and ever easy and freely available feedstock with minimum capital, thus lowering the need to regenerate the same from spent.

- The transport, storage and handling operations are very simple due to nonperishable nature of DCP.
- The HA extraction from DCP is also a *green process* wherein catalyst or any other auxiliary substances are not employed to add on to the cost, time, and chemicals, as compared to its other competitive methods presently dealt with.
- For the biosorption process, DCP is successfully employed without any pretreatment or modification process. Having porous and coarse nature settling is easy and hence it is efficient in decontamination.
- The disposal of DCP is easy as it can be sun dried and utilized as a landfill due to its minimum leaching nature.
- Being freely and easily available DCP has an edge over processed natural adsorbent considering their cost, time, and energy efficiency.

4.3.2 Weaknesses

- Biosorption mechanism of DCP still requires detailed studies due to its heterogeneous nature.
- Back extraction is difficult. Thus, research is required for stripping back of precious metals.

4.3.3 Opportunities

- Computer simulation will offer extremely powerful tools for designing.
- Metal recovery by different processes will prove Atom Economy.
- A need for R&D studies dealing more cogently with the molecular mechanisms, including such issues as binding sites, valence states, coordination chemistry, life cycle analysis and the speciation of metals in solution will surely optimize the Industrial Applications.

4.3.4 Challenges

- Direct Industrial Applications will be challenging.
- For a real application more research regarding direct impact of stability of partially or fully saturated DCP on nature is urgently needed.
- Eco-friendly Disposal Mechanism of DCP must be formulized for long term storage.

5. Conclusions

The Ancient Indian Epitome of Energy, DCP-unearthed ecofriendly resin has been scientifically studied to strongly echo its multispectral utility as a Humiresin, Green agrarian booster and Pseudo shale, sustaining WFE Nexus. The development of green and clean HA extraction process from DCP and its comparison with standard HA from IHSS holds great importance in the chapter. As our environmentalists and green chemists urge for *Reduce, Reuse, and Recycle* - the 3R's of Zero Waste Management theory, coined with Re-think, we can Re-balance the WFE Nexus and the sustainability of Life by employing DCP in various sectors as studied in this chapter. The utility of DCP will surely contribute to convert waste into wealth and refuse into resource, which is the ultimate goal of waste management strategy. This chapter tries to answer and express the great potential of DCP in the industry of Environmental Management, which can not only fix WFE Nexus but also the future and us.

Appendices and nomenclature

Appendix 1 – List of Figures

Appendix 2 – List of Tables

Author details

Hemlata K. Bagla
Department of Nuclear and Radiochemistry, Kishinchand Chellaram College, Mumbai, India

*Address all correspondence to: hemlata.bagla@kccollege.edu.in

References

[1] J. M. Nyika, Sustainable Ecosystem Management: Challenges and Solutions, Impacts of Climate Change on Agriculture and Aquaculture. 2021.

[2] V. Jenkins, "Sustainable Management of Natural Resources: Lessons from Wales," J. Environ. Law, vol. 30, no. 3, pp. 399-423, 2018.

[3] A. C. Smith, P. M. Berry, and P. A. Harrison, Sustainable Ecosystem Management. 2016.

[4] D. G. Blackmond, "The Origin of Biological Homochairality," Cold Spring Harb Perspect Biol., vol. 2, no. 5, 2010.

[5] M. A. Daei and M. Daei, "Humic first, A new theory on the origin of life," Geophys. Res. Abstr., vol. 18, no. EGU2016-11714, 2016.

[6] M. A. Daei, M. Daei, and B. Daei, "Humic First Theory: A New Theory on the Origin of Life," Geophys. Res. Abstr., vol. 19, no. EGU2017-2356, 2017.

[7] A. G. Zavarzinaa and G. A. Zavarzina, "Humic Substances in the Early Biosphere," Paleontol. Journal, vol. 47, no. 9, pp. 984-988, 2013.

[8] D. McCauley, "Soil Organic Matter Black Box," in Examining Alkaline Extraction and Humic Substances Research, 2020.

[9] P. Schmitt, A. Kettrup, D. Freitag, A. W. Garrison, and J. Fresenius, "Flocculation of humic substances with metal ions as followed by capillary zone electrophoresis.," Anal Chem., vol. 354, pp. 915-920, 1996.

[10] N. Senesi, "Humic Substances as Natural Nanoparticles Ubiquitous in the Environment," in Molecular Environmental Soil Science at the Interfaces in the Earth's Critical Zone., J. Xu and P. M. Huang, Eds. Springer Berlin Heidelberg, 2010.

[11] J. Weber, Y. Chen, and E. Jamroz, "Preface: Humic substances in the environment," J Soils Sediments, vol. 18, pp. 2665-2667, 2018.

[12] A. Muscolo, *S. Maria*, F. Ornella, V. Tugnoli, and N. Serenella, "The auxin-like activity of humic substances is related to membrane interactions in carrot cell cultures," J. Chem. Ecol., vol. 33, pp. 115-129, 2007.

[13] K. H. Tan, "Humic matter in soil and the environment: principles and controversies," in Boca Ranton: CRC Press 2nd edition, 2014, pp. 333-368.

[14] B. A. d. Melo, F. L. Motta, and M. H. Santana, "Humic acids: Structural properties & multiple functionalities for novel technological developments," Mater Sci Eng C Mater Biol Appl., vol. 5, no. 62, 2016.

[15] J. Kochany and W. Smith, "Application of humic substances in environmental remediation," in WM'01 Conference, 2001.

[16] J. Burlakovs, M. Kļaviņš, L. Osinska, and O. Purmalis, "The Impact of Humic Substances as Remediation Agents to the Speciation Forms of Metals in Soil," APCBEE Procedia, vol. 5, pp. 192-196, 2013.

[17] D. C. Olk et al., "Using Humic Fractions to Understand Natural Organic Matter Processes in Soil and Water: Selected Studies and Applications," J. Environ. Qual., vol. 48, pp. 1633-1643, 2019.

[18] J. Weber, "Humic Substances and their Role in the Environment," EC Agric., pp. 03-08, 2020.

[19] J. Allouche, C. Middleton, and D. Gyawali, "The Water–Food–Energy Nexus- Power," in Politics, and Justice, 2019, pp. 1-2.

[20] A. Hamidov and K. Helming, "Review Sustainability Considerations in Water–Energy–Food Nexus Research in Irrigated Agriculture," Sustainability, vol. 12, no. 15, p. 6274, 2020.

[21] P. Zhang et al., "Food-energy-water (FEW) nexus for urban sustainability: A comprehensive review," Resour. Conserv. Recycl., vol. 142, pp. 215-224, 2019.

[22] D. D. Konadu et al., "Not all low-carbon energy pathways are environmentally 'no-regrets' options," Glob. Environ. Chang., vol. 35, pp. 379-390, 2015.

[23] J. K. Seadon, "Sustainable waste management systems," J. Clean. Prod., vol. 18, no. 16-17, pp. 1639-1651, 2010.

[24] K. Ferrari, R. Gamberini, and B. Rimini, "The waste hierarchy: A strategic, tactical and operational approach for developing countries. The case study of Mozambique," Int. J. Sus. Dev. Plann., vol. 11, no. 5, pp. 759-770, 2016.

[25] X. Wang, M. Guo, R. M. Koppelaar, K. H. van Dam, C. P. Triantafyllidis, and N. Shah, "A Nexus Approach for Sustainable Urban Energy-Water-Waste Systems Planning and Operation," Environ. Sci. Technol., vol. 52, no. 5, pp. 3257-3266, 2018.

[26] H. Hettiarachchi and C. Kshourad, "Promoting Waste-to-Energy: Nexus Thinking, Policy Instruments, and Implications for the Environment," Waste Treat. Process. Energy Gener., vol. 9, pp. 163-184, 2019.

[27] B. Feng, K.H.VanDam, M. Guo, N. Shah, S. Passmore, and X. Wang, "Planning of Food-Energy-Water-Waste (FEW2) nexus for sustainable development," BMC Chem Eng., vol. 2, no. 4, 2020.

[28] K. M. Goh, "Carbon sequestration and stabilization in soils: Implications for soil productivity and climate change," Soil Sci. Plant Nutr., vol. 50, no. 4, pp. 467-476, 2004.

[29] J. Frouz et al., "Interactions between soil development, vegetation and soil fauna during spontaneous succession in post mining sites," Eur. J. Soil Biol., vol. 44, no. 1, pp. 109-121, 2008.

[30] W. Xinyu et al., "Effects of Litter fall on the Accumulation of Extracted Soil Humic Substances in Subalpine Forests," Front. Plant Sci., vol. 11, 2020.

[31] S. Dou, Y. Tardy, J. J. Zhang, L. Kai, S. Q. Yu, and L. F. Ping, "Thermo dynamic stability of humic acid and fulvic acid in soil and its driving factors," Acta Pedol. Sin, vol. 47, pp. 71-76, 2010.

[32] W. G. Botero, L. C. Oliveira, A. D. M. Cavagis, A. H. Rosa, J. C. Rocha, and A. Santos, "Influence of the extractant on the complexing capacity of humic substances from peat for macro and micronutrients using continuous flow: agricultural application and environmental impacts," J. Braz. Chem. Soc., pp. 12-24, 2013.

[33] S. H. Baba, M. S. Bhat, and M. A. Bhat, "Zero Waste: A Sustainable Approach for Waste Management," 2020.

[34] C. Feller, E. Blanchart, and A. Herbillon, "The Importance of French Tropical Research in the Development of Pedology," SSSAJ, vol. 72, no. 5, 2008.

[35] K. Cimrin, O. Turkmen, M. Turan, and B. Tuncer, "Phosphorus and humic acid application alleviate salinity stress of pepper seedling," African J. Biotechnol., vol. 9, no. 36, 2010.

[36] H. Feng et al., "Carbonized Cow Dung as a High Performance and Low Cost Anode Material for Bio electrochemical Systems," Front Microbiol., vol. 30, no. 9, p. 2760, 2018.

[37] S. Kumar, H. D. Kumar, and . K. G Babu, “A study on the electricity generation from the cow dung using microbial fuel cell,” J.Biochem Tech, vol. 3, no. 4, pp. 442-447, 2012.

[38] K. K. Gupta, K. R. Aneja, and D. Rana, “Current status of cow dung as a bioresource for sustainable development,” Bioresour. Bioprocess, vol. 3, no. 28, 2016.

[39] N. S. Barot and H. K. Bagla, “Eco-friendly waste water treatment by cow dung powder (Adsorption studies of Cr(III), Cr(VI) and Cd(II) using tracer technique),” Desalin. Water Treat., vol. 38, no. 1-3, pp. 104-113, 2012.

[40] N. D. Bhatt and H. K. Bagla, “Sustainable remediation of Hg(II) from wastewater by combo humiresin-dry cow dung powder,” J. Environ. Biotechnol. Res., vol. 6, no. 1, pp. 168-178, 2017.

[41] S. Shaikh and H. Bagla, “Comparative study of 110mAg(I) removal from aqueous media by humic substances,” J Radioanal Nucl Chem, 2019.

[42] S. A. Shaikh, “To study biosorptive profile of Ag(I)-110m and Zn(II)-65 employing humiresins,” University of Mumbai, 2019.

[43] S. Sayed and H. Bagla, “Sorption Studies of Radionuclides from Simulated Low Level Waste using Green Biosorbent,” Res. J. Chem. Environ., vol. 25, no. 1, pp. 1-6, 2021.

[44] A. Khan and H. Bagla, “Application of tracer technique in remediation of Sr(II) from simulated low level radioactive waste,” J. Radioanal. Nucl. Chem., vol. 322, Apr. 2019.

[45] N. S. Barot, R. P. Khilnani, and H. K. Bagla, “Biosorptive profile of synthetic and natural humiresin for the remediation of metallic water pollutants,” J. Radioanal. Nucl. Chem., vol. 302, no. 2, pp. 951-959, 2014.

[46] O. Obire, R. N. Okigbo, and C. F. Ojim, “Fungal population and diversity in partially digested cellulose from the abomasum of beef cows,” J. Agric. Technol., vol. 6, no. 4, pp. 783-792, 2010.

[47] K. Yokoyama and N. Miyauchi, “Decomposition of organic matter in cow dung colonized by Onthophagus lenzii 249 Harold (Coleoptera: Scarabaeidae),” Edaphologia, vol. 47, pp. 33-39, 1991.

[48] Y. L. Nene, “Seed health in ancient and medieval history and its relevance to present day agriculture,” Asian-Agri-History, vol. 3, pp. 157-184, 1999.

[49] K. C. Teo and S. M. Teoh, “Preliminary biological screening of microbes isolated from cow dung in Kampar,” African J. Biotechnol., vol. 10, no. 9, pp. 1640-1645, 2011.

[50] H. Boricha and F. MH, “Pseudomonas plecoglossicida as a novel organism for the bioremediation of cypermethrin,” Biol. Med., vol. 1, Jan. 2009.

[51] G. B. P. P. A. Joshi, “Screening of petroleum degrading bacteria from cow dung,” Res. J. Agric. Sci., vol. 2, no. 1, pp. 69-71, 2011.

[52] M. Swain, R. Ray, and C. Nautiyal, “Biocontrol efficacy of Bacillus subtilis strains isolated from cow dung against postharvest yam (Discora rotundata L.),” Cur. Microbiol ., vol. 57, pp. 407-411, 2008.

[53] O. O. Akinde SB, “Aerobic heterotrophic bacteria and petroleum-utilising bacteria from cow dung and poultry manure.,” World J Microbiol Biotechnol, vol. 24, no. 9, pp. 1999-2002, 2008.

[54] A. Wang et al., "Biomass-derived porous carbon highly efficient for removal of Pb(II) and Cd(II)," Green Energy Environ., vol. 4, no. 4, pp. 414-423, 2019.

[55] T. Satoh, "Organo-mineral complex status in soils: Thermal nature of organo-mineral complex particles and their humic substances," Soil Sci. Plant Nutr., vol. 30, no. 1, pp. 95-104, 1984.

[56] *L. campanella* and M. Tomassetti, "Thermogravimetric and IR analysis of different extracts of humic substances," Thermochim. Acta, vol. 170, pp. 67-80, 1990.

[57] P. Ioselis, Y. Rubinsztain, R. Ikan, Z. Aizenshtat, and M. Frenkel, "Thermal characterization of natural and synthetic humic substances," Org. Geochem., vol. 8, pp. 95-101, 1985.

[58] "Analytical spectroscopy," 1994.

[59] E. Abakumov, E. Lodygin, and V. Tomashunas, "13C NMR and ESR Characterization of Humic Substances Isolated from Soils of Two Siberian Arctic Islands," Int. J. Ecol., 2015.

[60] M. Tatzber et al., "FTIR-spectroscopic characterization of humic acids and humin fractions obtained by advanced NaOH, Na4P2O7, and Na2CO3 extraction procedures," vol. 170, no. 4, pp. 522-529, 2007.

[61] H. K. Bagla, "Extraction of humic acid from biological matrix – dry cow dung powder AU - Barot, N.S.," Green Chem. Lett. Rev., vol. 2, no. 4, pp. 217-221, Dec. 2009.

Chapter 3

Co-Composting of Various Residual Organic Waste and Olive Mill Wastewater for Organic Soil Amendments

Wafa Hassen, Bilel Hassen, Rim Werhani, Yassine Hidri and Abdennaceur Hassen

Abstract

The valorization of different organic residues like municipal solid wastes, sewage sludge and olive mill wastewater is becoming more and more worrying in the different modern communities and is becoming relevant and crucial in terms of environmental preservation. The choice of the treatment technique should not be only from the point of view of economic profitability but, above all, must consider the efficiency of the treatment method. Thus, an attempt to remove polyphenols from olive mill wastewater would have a double interest: on the one hand, to solve a major environmental problem and to recover and valorize the olive mill wastewater for advanced applications in food processing and soil amendments. It is also interesting to think of associating two harmful wastes by co-composting such as sewage sludge-vegetable gardens, sewage sludge-municipal solid waste, and green wastes-olive mill wastewater..., to get a mixed compost of good physical–chemical and biological qualities useful for agricultural soil fertilization. Finally, in order to be more practical, we will describe specifically in this chapter a new variant of composting and co-composting technology intended for waste treatment that is very simple, inexpensive and easy to implement.

Keywords: Valorization, Wastes, Olive mill wastewater, Compost, Soil fertilization

1. Introduction

Compost makes up a stable, hygienized and humus-rich product resulting from the mixing of various municipal, plant or animal residues, gradually fermented to ensure the decomposition of organic matter (OM), and used as a fertiliser, amendment or growing medium. Thus, compost is the resulting product of a complex microbial process of decomposition and transformation of biodegradable organic residues. This process operated under varied microbial communities that develop and grow in aerobic conditions [1–3]. Fermentable organic wastes are coming in many forms and with different proportions and accessibility to microorganisms. The raw materials used in composting are much diversified, we can cite as examples raw manure, bedding, feed residues, straw, various crop residues, olive pomace and varied products from agro-food industries [4].

Composting is a biochemically continuous phenomenon of organic matter mineralisation or oxidation in the presence of oxygen. The mineralisation or oxidation is achieved by microorganisms that use oxygen from the air and organic carbon for all their all-metabolite biosynthesis. To make the degradation or the oxidation easy, two operations may be implemented. These two operations are always considered as optional. First, the waste can be sorted to separate the fermentable fraction from the non-recyclable one. Second, the waste may be mechanically shredded to improve the structure of the waste mass; thus, on the one hand, the waste shredding facilitated the biodegradation, reduced the treatment time and make handling more pleasant; and the second hand to make homogenisation of the waste mass easy by allowing a uniform distribution of the different waste components. Once the residual substrate has been prepared, fermentation that resides at the heart of the process, is started. This fermentation leads to a rapid decomposition of easily biodegradable organic matter that generates some fewer complex molecules. Subsequently, the substrate biodegradation leads to slower maturation. These steps are commonly known as the processes of humification and stabilisation of the compost [5]. Once the compost has reached maturity, it will undergo screening and sieving. This operation is allowed to give two products: a commercial product known as compost and a refusal product to refine and/or to landfill. All these mentioned operations are well conducted in all common composting processes. Despite, some precise differences could be existed and lie in the location of the screening phase and the choice of the fermentation system. The originality aim of this book chapter is to describe, in the composting plant implemented under semi-arid pedoclimatic conditions, a description of new and simple process of composting and co-composting of municipal solid wastes, olive mill wastewater, waste farm and garden cutting, straw and sewage sludge. All these residual materials resulting from the main human activities commonly known at the present state in modern societies, with especially the olive mill wastewater resulting from oleiocultural activities, will be considered in this a new process of composting description, mainly characterized as simple to implement and to monitor [6, 7].

Indeed, the process is directed in two principal successive steps: step of prefermentation (uncontrolled fermentation) and step of maturation (controlled fermentation), respectively.

2. General composting process operation

Composting is a complex bio-physical–chemical operation that comprises biodegradation of organic waste under controlled conditions of temperature, humidity and aeration. Two important following one another biophysical phenomenon could include the common composting process. The first process brings the organic residues to the state of fresh compost. An intense aerobic degradation concern essentially the decomposition of fresh organic matter at a high temperature of 50–70°C under the action of thermophilic bacteria; while the second process is done by a less sustained degradation mainly achieved by mesophilic bacteria. These bacteria transform the fresh compost into a mature compost, rich in humus. This maturation phenomenon, which takes place at lower temperatures of 35–45°C, leads to the biosynthesis of humic compounds by fungi.

The composting process of organic residues takes place in three distinct and important phases. First, the temperature rises rapidly to around 40°C or 45°C following the respiration of aerobic mesophilic microorganisms; in parallel, the most degradable compounds such as sugars and starch are consumed. Second, the temperature rises progressively to around 60°C or 70°C and the mesophilic microorganisms will be replaced by thermophilic ones, anaerobic fermentation by aeration of the waste mass must be avoided; pathogens, parasites and weed seeds will be destroyed by the temperature. Third, the degradation is complete when aeration no longer increases the temperature, the amount of material easily operational by the microflora becomes scarce and the biosynthesis of humic compounds becomes predominant, and at the end the thermophilic species in favour of more common species disappears and appearance of new mesophilic species [8].

3. Physical-chemical parameters monitored during the composting process

The principal physical–chemical parameters of the process monitoring are summarised in the parameters that condition the good development and progress of microbiological activities, and their monitoring is essential to test the effective conduct and behaviour of the composting process [9, 10]. This is achieved by optimising nutrient supply and regulating pH, temperature, water content and aeration conditions.

3.1 Grain size of wastes

The waste particle size is an important parameter to consider in the composting process since (i) it determines the size and volume of the pores created by the arrangement of the particles in the waste matrix, (ii) it is involved in increasing the specific surface area of the raw organic matter, (iii) it facilitates the diffusion of oxygen inside the compost waste matrix, thus allowing homogenisation of the waste, and at last (iv) it is the site of main microbiological activities that take place on the surface of the organic particles.

3.2 Interstices oxygen rate

This parameter shows the real proportion of oxygen in the interstices of the waste mass. It is critical to the oxidation of the organic matter, and directly related to the size, humidity and aeration of the waste during composting. Oxygen requirements decrease along composting whether they are proportionate to the organic matter gradually disappearing over the mineralisation process. However, maintaining and preserving good aeration avoids the start of an anaerobic process that could induce the generation of malodorous compounds. Moisture in the waste mass always interacts negatively with the system aeration. The supply of oxygen allows the drop in humidity. If this humidity is high, a probable temperature rise will take place leading to a significant improvement in the substrate mass homogeneity. The minimum threshold of oxygen needed to maintain aerobic conditions is of the order of 5% as reported by Jammes [9].

3.3 Prevailing humidity in the waste mass

Humidity is both a raw material-related parameter and a monitoring parameter. It hosts the development of the microbial flora within the compost. The optimal water content during composting is around 60%. However, high water content promotes anaerobic fermentation. If the water content exceeds 70%, the water fills the voids and space, making oxygen exchange very difficult. On the other hand, if this prevailing humidity drops below 20%, the decomposition of organic matter will be inhibited.

The quantity of water lost by vaporisation during the release of heat exceeds those formed during the reaction of oxidation; therefore, to compensate this lessening, it is tolerated and needed watering materials during composting. Although it is difficult to determine the volume to be added, water can be added as long as no runoff appears under the pile of waste or the waste mass.

3.4 C/N ratio of the waste mass

The C/N ratio of the waste mass qualifies the biodegradability of organic waste by ensuring the trophic balance necessary for the flora optimal development. Suitability of fermentable waste for composting is determined by the C/N ratio of the mixture of their various constituents (fermentable, paper and cardboard). Putrescible materials whose C/N are of the order of 15, are substrates easily biodegradable, while paperboards, with C/N ranging from 60 to 107, are substrates hardly biodegradable.

During aerobic fermentation, microorganisms consume carbon 15 to 30 times more than nitrogen. The initial C/N ratio is around 30 to 35, while that of the final product is less than 15. Sometimes the C/N ratio of waste can be so low that it is unsuitable for composting. This can be remedied by adding a specific substrate with a high C/N ratio, which brings the initial C/N value back towards the optimum [9]. If the initial ratio is less than 30, nitrogen losses are accompanied by the odour nuisance. If the ratio is low, ammonia nitrogen losses may reduce pH. **Tables 1** and **2** showed examples of the C/N values of some compostable materials.

3.5 Temperature developed inside the waste mass

The increase in temperature is caused by microbiological activity. During the degradation of organic matter, there is energy initially contained in the chemical bonds of the constituent molecules, which are released, part of which is recovered by the metabolism of microorganisms, and the other part is dissipated into the atmosphere. Therefore, the minimum temperature is necessary for degradation and the evolution of the temperature during composting is allowed distinguishing four successive distinct phases [1].

Categories	Season	Average C/N
Fermentables	Spring	18.3
Paper-Cardboard	Spring	63.4
Fermentables	Summer	13.1
Paper-Cardboard	Summer	59.2
Fermentables	Autumn	15.6
Paper-Cardboard	Autumn	107.5
Fermentables	Winter	15.2
Paper-Cardboard	Winter	79.1

Table 1.
Average C/N ratios of some fermentable components.

	Municipal wastes	Cattle manure	Sewage sludge	Olive pomace
C/N	< 20	20	11	49.3

Table 2.
Average C/N ratios of main fermentable components.

3.5.1 The mesophilic phase

Mesophilic microorganisms (especially bacteria and fungi) invade the raw material, so their activity causes a rise in temperature (from 10 to 15°C to 30–40°C), a significant release of CO_2 and subsequently a decrease in the C/N ratio and acidification.

3.5.2 The thermophilic phase

During this phase, the temperature reaches 60 to 70°C, values to which only thermo-tolerant microorganisms like actinomycetes and thermophilic bacteria could remain in operation in this very hostile environment, and the degradation activity of resistant fungi will be stopped. Along this hostile phase, the nitrogen mineralized as NH4+ will be lost as NH_3. This hostile environment takes place specifically inside the waste mass centre, hence leading to the need to turn over the waste mass for ensuring homogeneous and disinfected products.

3.5.3 The cooling phase

This phase is followed by the operation of turning and watering the mass of waste being composted. Thus, it is mainly characterised by the reappearance of ambient temperature and mesophilic microorganisms that decompose materials remained intact during the previous phase and nitrogen of some complex components.

3.5.4 The maturation phase

The microbiological activity regresses considerably during this phase, and the waste receives new colonisers that are the macrofauna, particularly earthworms. The organic matter becomes stabilised and humified compared to their initial state. It should be pointed out that a temperature above 70°C should be avoided, as it leads to extreme drying, a very significant loss of material and even a halt in the process by the microflora destruction.

3.6 pH or acidity degree

The hydrogen potential known as pH is a measure of the chemical activity of protons or hydrogen ions in solution inside each medium. The pH largely influences the development of the microflora responsible for the waste decomposition. Its value is determined and imposed by the raw material used, but varies according to the progress of the composting process. The pH monitoring is mandatory since it provides information on the different phases of the process. The optimal pH prevailing in the waste mass during the composting is on average between 6 and 8.

3.7 Undesirable and non-biodegradable waste products

The composition of municipal waste may show undesirable non-biodegradable elements that could affect the process and the ultimate product quality. These elements may be notable as packaging and special wastes, rich in metallic elements. Therefore, these undesirable materials should be separated by sorting. The operation of sorting can be made at the domestic level known as source sorting prevailing and well distributed in very advanced modern societies and/or in composting sites.

4. Description and implementation of a new composting process

We are prompted in this paragraph to describe a suite of a new composting process, traditional, very easy to implement, economical, but it is a time and space-consuming process on the plant composting platform. The process is simply summed up in two successive stages, recognized as the pre-fermentation and maturation stages.

As earlier said several times, the composting process at the microbial level involves many interrelated factors, mainly metabolic heat generation, temperature, ventilation as O2 input, moisture content and nutrients. Temperature profiles registered during the two steps of pre-fermentation and maturation are shown in **Figure 1** in two experimental wastes windrows W1 and W2 the three classical temperature steps of composting, including the mesophilic, thermophilic and cooling phases, respectively (**Figure 1**).

The step of maturation with around 60 days appeared to be as less active as compared to the first step of pre-fermentation with around 90 days since the values of temperature, registered during the second step of maturation, are less important than those to be recorded during the first step. Differences in temperature averaged 15°C. This result is mainly related to differences in the availability of easily decomposable organic matter content as nutrients in the organic fraction. So, composted materials used during the first step of pre-fermentation seem richer in these easily decomposable organic elements as compared to those of the second step of maturation.

Analytical investigation of some key monitoring biological parameters of composting such as dehydrogenase activity [11], microbial biomasses C (BC) and N (BN), extracted total DNA content and microbial diversity in waste masses during composting were examined. So, the dehydrogenase activity is studied during the composting process for two major reasons: the first concern the follow-up of the biodegradation level of substrates, and the second because it represents a reliable indicator of the stability and maturity of the finished product. The dehydrogenase

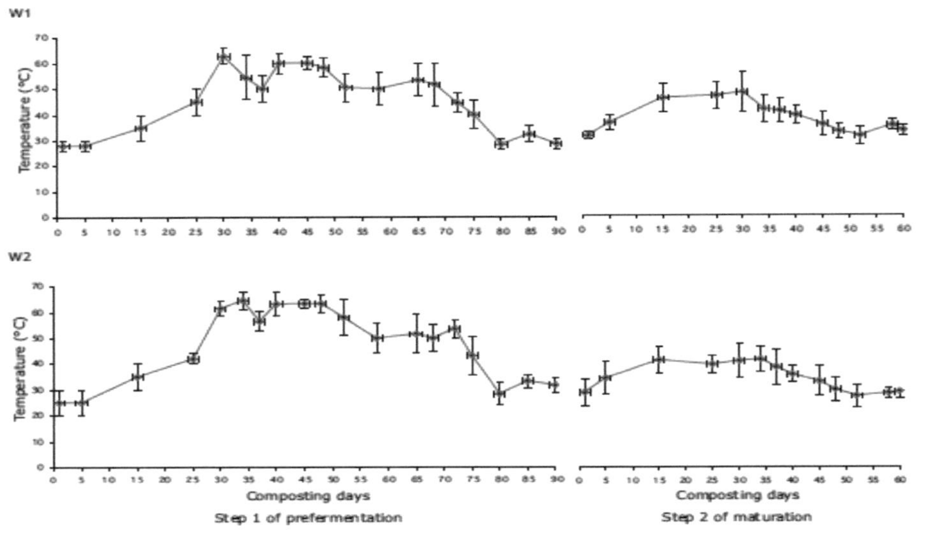

Figure 1.
Temperature changes in the two windrows W1 and W2 during the two steps of composting. (i) The first windrow W1 constituted with 100% of municipal solid wastes, (ii) The second windrow W2 composed by weight of 60% of municipal solid wastes and 40% of dried stabilized sewage sludge, and (iii) The form of windrows of 12 × 6 × 2.5 m (length x width x height, respectively).

activity showed a net increase between 2 and 0 and 15–35 days through the pre-fermentation and maturation steps, respectively. A good positive correlation is observed between the values of dehydrogenase and of temperature registered ($r = 0.96$, $P < 0.01$). These values of dehydrogenase activity fluctuated between 3.2–5.84 and 1.7–5.66 mg TPF/g of waste dry weight/24 h during the step of pre-fermentation in the two windrows W1 and W2, respectively (**Figure 2**). By the same, these values varied between 3.6–4.9 and 1.96–3.22 mg TPF/g of waste dry weight/24 h during the step of maturation in the two windrows W1 and W2, respectively. So variation of the dehydrogenase activity appeared very narrow (around 1.2 mg TPF/g of waste dry weight/24 h on average for W1 and W2) during the second step of composting and showed a certain homogeneity and consistency of the waste materials used in this second step of maturation; and in the opposite, a high heterogeneity and assortment of the waste used during the first step of pre-fermentation since variation of dehydrogenase values generally recorded were relatively large (around 2.64 and 4 mg TPF/g of waste dry weight/24 h for W1 and W2, respectively).

On the other hand, it is important to mention that the dehydrogenase activity values appeared slightly higher in the windrow free of sludge than in the one with sewage sludge. The microbial biomasses C and N behavior biochemical transformations in waste materials, and BC/BN ratio is closely related to the changes of the microbial population during composting. BC/BN ratio values usually showed a net increase during the thermophilic phase, 30–70 and 20–50 days, for the steps of pre-fermentation and maturation, respectively (**Figure 2**). The BC/BN ratio values are on average around 4 or 6 and 3–4 during the first and second steps of composting, respectively. The concept of the microbial biomass inventory regards the microorganisms as only one and a single entity. The evolution of BC/BN ratio translates a microbial diversity, with phases where they are the bacteria and the actinobacteria which are dominant, and others where the fungi prevail; a net increase of this ratio is synonymous of a good microbial activity [12–15].

Microbial total DNA extracted from composting materials and followed during all the two steps of the composting process showed a net variation over time, and

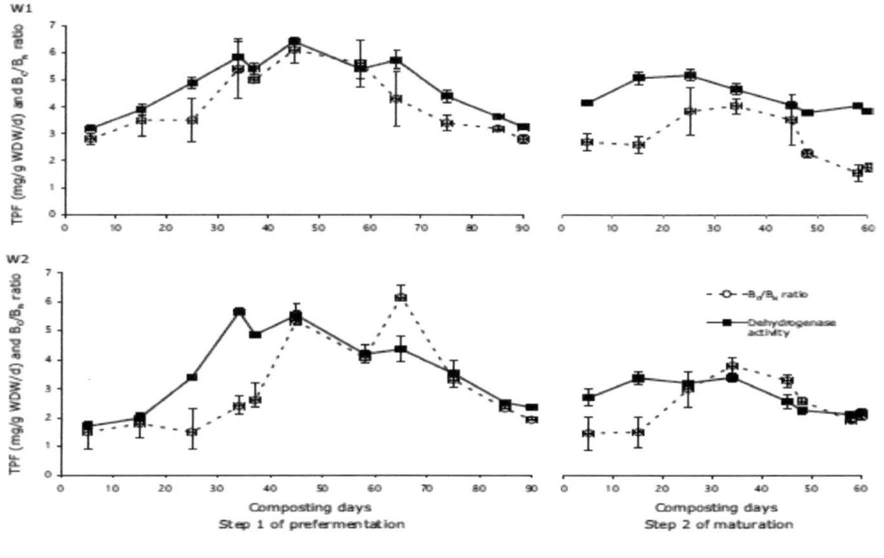

Figure 2.
Dehydrogenase activity and BC/BN ratio changes in two windrows W1 and W2 during the two steps of composting.

revealed a good parallel increase with the temperature progress inside the waste materials (**Figure 3**). This increase varied between 13.2 and 26.1 µg of total DNA per g dry weight. The lowest values of DNA are observed at the start and the end of each step of the process (around 13.2 µg of total DNA per g dry weight) and the highest values are always registered during each thermophilic phase (around 26 µg of total DNA per g dry weight). The ratio of absorbance at 260 and 280 nm is usually used to assess the purity of DNA and RNA. A ratio of ~1.8 is generally accepted as 'pure' for DNA; a ratio of ~2.0 is generally accepted as 'pure' for RNA. If the ratio is appreciably lower in either case, it may indicate the presence of protein, phenol or other contaminants that absorb strongly at or near 280 nm. The determination of A260/A230 and A260/A280 ratios usually defined as a coefficient of purity of DNA. For compost DNA showed a significantly lower value (0.96 and 1.2) than those for DNA solutions of pure cultures (1.57 and 1.89) showing that compost DNA was coextracted with humic and protein compounds, respectively.

Therefore, the investigation concerning microbial diversity in the composted wastes allowed to assess the changes in microbial diversity that could occur during the two steps of the process by DNA extraction and polymerase chain reaction (PCR). Main investigation and results showed that along the first process step, there is a net variation in microbial diversity. This diversity appeared very rich and obvious in case of windrow W2 added with sewage sludge. The principal characteristic of the first step of pre-composting or pre-fermentation consists in subjecting directly to fermentation raw wastes, without sorting and crushing.

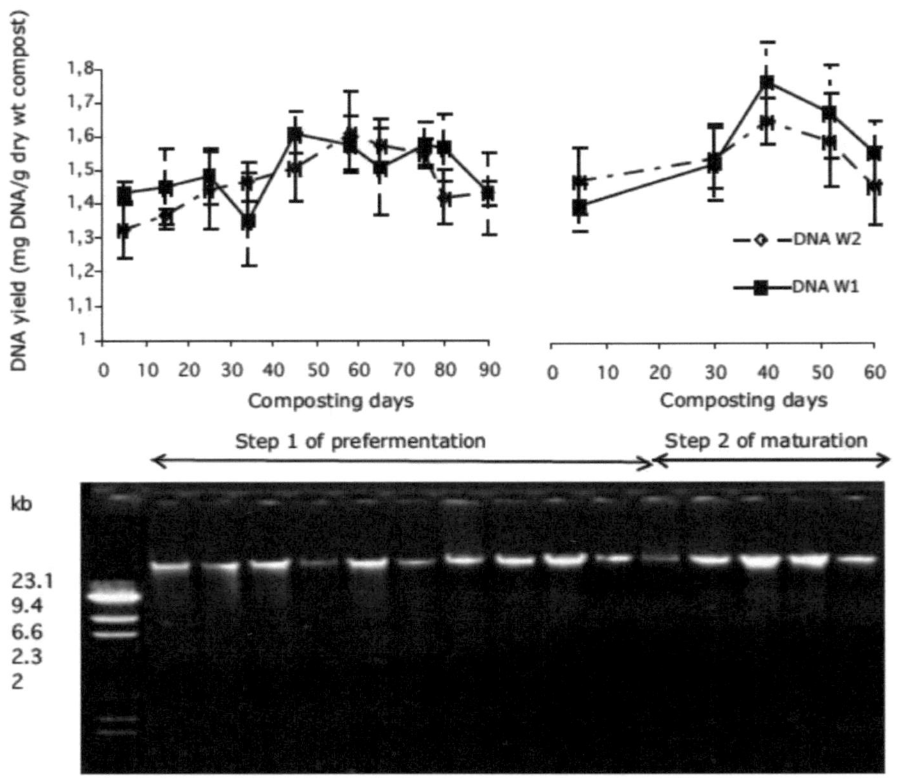

Figure 3.
Total compost DNA contents and agarose gel electrophoresis of total DNA extracted from composting materials (W1) during the two steps of composting. First lane: HindIII-cut bacteriophage lambda molecular size markers (1 mg), step 1 and step 2 lanes: DNA extracted from waste of W1.

The usual processes suppose the use of a pre-sorted and crushed raw material. In the case of the present composting technique adopted in this process, raw wastes not sorted upstream are very heterogeneous, and then the separation of organic material is practically impossible. Thus, the operation of pre-fermentation supports accessibility to this organic material. During this first step and earlier evoked, the rises of temperature appeared important and reach the 60–65°C inside the two windrow W1 and W2 of waste masses. Also, the degradation of organic material makes possible the decrease of plastic bags and the release of organic matter that it is frequently locked up there.

This pre-fermentation is preferred done into two pyramidal windrows forms and pile of waste should be often covered with a layer of fresh sewage sludge (approximately 30% of total waste weight). In fact, this layer of sludge participates efficaciously to minimize the fire hazard and to avoid the take-off of certain particles of waste, such as plastic, paper, and many others.

The optimal duration of this phase of pre-fermentation was previously studied, and evaluated to three months on average. Consequently, a maximum organic material recovery is ensured during a minimum time interval.

The second step of the process of controlled fermentation (step of maturation) is characterized by a degradation of the organic materials by microorganisms. The purpose of management of this second step of composting is to ensure favorable conditions in order to force the microbiological activity. The procedure of microbes during this second step of composting appeared on average less active than the one registered during the first step of pre-fermentation. All parameters considered in the present paragraph are in favor of this conclusion. We could explain this result by several factors mainly by the decrease or exhaustion and depletion of readily degradable compounds in the mass of waste materials.

This chapter review confirms amongst others that temperature is the main parameter to consider in composting. This important factor conditions the functioning of all the other parameters, specifically the microbial activity. As a result, the level of temperature conditions and depends on the type of microorganisms operating during composting. So mesophilic microbes dominate at the beginning and at the end of each step of the composting process; on the other hand, thermophilic microbes control the important step of composting, the thermophilic phase.

Finally, this paragraph reviews a new and simple composting process tested and used in an industrial plant for compost production under semi-arid pedo-climatic condition, and some important chemical, physical and biological monitoring parameters should be investigated in order to understand and to master the general process of composting of various solid residues. A molecular detection procedure, using ribosomal intergenic spacer analysis, was tested for microbial community diversity assessment. At last, analysis of compost microbial communities is one of the challenging areas of research due to the enormous complexity of biodiversity caused by the heterogeneity of the physical and chemical structure of compost environments [16, 17].

5. Co-composting process of municipal solid wastes, green cutting waste straw and olive mill wastewater

As earlier evoked, olive production represents one of the oldest agricultural activities in the Mediterranean basin. For these countries, the production of olive oil is an economic fortune transmitted over several generations. However, it has the disadvantage of generating huge quantities of by-products with a complex organic fraction and high chemical oxygen demand (COD). Indeed, 100 kg of olives produce on average 35 kg of pomace and 100 liters of vegetable oil [18]. Thus, these residues are commonly considered as an important polluting industrial waste.

Currently, with the promotion of the beneficial virtues of olive oil for human health, its demand continues to increase and consequently production is constantly growing at the expense of the environment. The discharge of effluents from olive oil mills has until now been a challenging ecological concern in the Mediterranean region. Oil plants furnished with modern equipment produce large quantities of 6 to 7 million tons/year that could reach 80–110% of the initial batch of olives, while with traditional devices, the production of olive oil mills is 50%. Given these excessive volumes of waste, treatment is essential to reduce the environmental impact. The problem posed by olive oil mills discharges resides mainly in their high polyphenol content that could reach 18–125 mg/g according to the variety of olives, the level of production, the period of olive picking and extraction.

From the preceding, it is clear that an attempt to remove polyphenols from the olive oil mills also known as olive mill wastewater would have a double interest: on the one hand, to solve a major environmental problem and, on the other hand, to recover and valorize the olive mill wastewater for later applications in agro-food.

It is also interesting to think of associating some other kinds of residual wastes that could be disruptive and sometimes harmful to the natural environment like gardening and crop residues, straw, green waste and dried sewage sludge for an attempt at recovery in terms of co-composting in order to evaluate the quality of the compost produced.

In this paragraph, a variant of the co-composting technique has been presented and described: olive mill wastewater with municipal solid waste, or olive mill wastewater with garden and cutting wastes, olive mill wastewater with dried sewage sludge.

5.1 Setting up waste windrows

As mentioned above, the technique adopted for this application is that of windrow co-composting. The windrows will be set up as the waste arrives at the composting plant. Care will be taken to ensure that the date of termination of the windrow placement is recorded. Also, it should be noted that the raw material collected has not been sorted or crushed before. The windrows were deposited on the controlled fermentation platform in pyramidal form (L x W x H = 12 × 6 × 2.5 m) (**Figure 4**). Co-composting processes are commonly carried out with variants of organic residues such as straw, cutting and garden waste, and sludge stabilized by drying in a natural sunny bed. The straw was chosen because of its structure that can absorb a large amount of olive mill wastewater and it is for the same purpose that the cutting waste was chosen. As for the WWTP sludge, it was chosen with the aim of looking for an efficient variant of recovery of this type of environmentally harmful waste. Main parameters of the composting process will be monitored in situ or laboratory measurements. Daily monitoring of the temperature considered as a key parameter of the composting process will be carried out using a compost thermometer (probe). Thus, a total of 9 × 3 measurements will be taken: three at the mid-height of the windrow (50 cm), three others at 150 cm from the bottom and lastly three measurements at 230 cm from the top and at three different points A, B and C of the windrow surface mass. The temperature of the windrow at a specific point will be taken as the average value of the temperatures at the different measuring points. Windrow turning made necessary after the temperature elevation within the waste mass allows a heterogeneous waste at the start of the process to be progressively homogenized; the watering will maintain a suitable humidity for the development of the different macro and microorganisms.

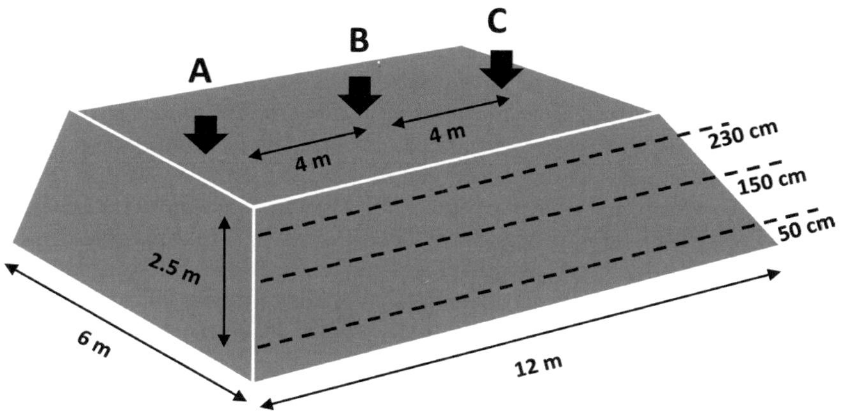

Figure 4.
Diagram showing the different levels of temperature measurement and waste sampling in a waste windrow.

The various possible treatment combinations could be conducted as follows: A1: Windrow composed of green waste without olive mill wastewater; A2: Windrow composed of green waste +15% olive mill wastewater; A3: Windrow composed of green waste +30% olive mill wastewater. A4: Windrow composed of premature compost without olive mill wastewater; A5: Windrow composed of premature compost +15% olive mill wastewater; A6: Windrow composed of premature compost +30% olive mill wastewater; A7: Windrow composed of straw + waste cuttings; A8: Windrow composed of straw + waste cuttings with 15% olive mill wastewater; A9: Windrow composed of straw + waste +30% olive mill wastewater. A10: WWTP sludge without olive mill wastewater; A11: WWTP sludge with 15% olive mill wastewater; A12: WWTP sludge with 30% olive mill wastewater, A13: Straws without olive mill wastewater; A14: Straws with 15% olive mill wastewater; A15: Straws with 30% olive mill wastewater; A16: Saturated straws. All these treatment combinations will be carried out in order to choose the right treatment for the different waste that goes with their rate and nature, and above all the quality of the finished product, i.e. the finished compost.

5.2 Quality of the finished products

It is important to mention that the olive mill wastewater addition to the sledges results in a less aerated compact waste structure of the windrow. This reduction of the interstitial space leads to a reduction in aeration at the level of the windrow, accompanied necessarily by a reduction in temperature at the level of the WWTP sludge-based windrows. At the level of these windrows, the temperature does not generally exceed 60°C. Therefore, monitoring of pathogens must be considered in order to ensure the sanitary quality of the finished product of compost.

However, the temperature changes in the straw-based windrows such A14, A15 and A16 showed high temperature values exceedingly largely 60°C, indicating active degradation. However, the windrow that has not received the olive mill wastewater, the temperature values to be recorded during the cycle of composting rates remains relatively low, this may be due to the fact that the straw has low enough moisture content for microbial growth, and an olive mill wastewater will be needed to correct the composition of the windrow in terms of moisture and nutrients.

Also, significant temperature increases will result in peaks between 50°C and 65°C. However, these thermophilic phases are interrupted by falls caused by overturning or naturally by precipitation. These waste turning is necessary to ensure aeration of the windrow and exudation of leachate from the windrows.

A global reading of the different temperature changes in the different variant of windrows could show that the composting cycle begins with a mesophilic phase where the temperature progresses rapidly from 40 to 45°C, just after a few days (5 to 6 days) of windrow implementation. Later and during the second thermophilic phase of composting, the temperature rise, which can last a month and a half, ensures pasteurization resulting in compost of good microbiological quality.

On the other hand, moisture is an essential factor in controlling the level of progress of the composting process. Here, humidity could vary between 20% and 85% with losses of water by evaporation and runoff or gains of water returned following rainfall in the absence of a cover. However, excessive humidity can interfere with the normal progress of the composting process, which may justify the aberrations that can sometimes be noted during temperature evolution.

Green waste and cutting windrows can have relatively higher moisture values of 65–88% than other straw windrows (A13, A14, A15 and A16). This may be attributed to their remarkable volumes that could create vacuums within the windrow.

However, moisture is less important for the sludge windrows. The small grain size of the sludge causes the windrows to become stacked. For straw windrows, low moisture values were recorded, suggesting that straw is a good absorbent, but significant leaching could be occurred throughout the composting cycle.

As mentioned earlier, the C/N ratio measures the relative proportions of carbon and nitrogen, nutrients essential to the life of microorganisms. Carbon and nitrogen measurements were taken at the beginning and end of the composting cycle. The highest ratios were recorded for straw swaths (A13, A14, A15 and A16) and cuttings (A7, A8, A9). This is because of their high carbon content, which is one molecule that is difficult to biodegrade. Hence its persistence in large proportion until the end of the composting cycle.

As for the windrows (A7, A8, A9, A13, A14, A15 and A16), respectively of cuttings and straw waste, relatively low nitrogen values of 0.30; 0.60; 1.00; 0.40; 0.80; 0.90 and 1.20% will be recorded at the end of the composting cycle.

As a result, C/N ratios remain fairly high, fluctuating between 124 and 139. This fairly high C/N ratio of the cuttings and straw waste windrows suggests that the resulting compost may be immature. It is well shown and known in the literature on this specific topic that the C/N ratio is considered an important indicator of maturity of organic matter in aerobic fermentation.

Green waste windrows represent the lowest final carbon values. This type of result automatically suggests compost stability, in terms of biodegradation speaking. In fact, green waste is essentially composed of cellulose, a component that is easily biodegradable. This specific composition means that the C/N ratios are always low. Indeed, a high initial C/N ratio favors nitrogen immobilization and a low ratio favors mineralization. This previous character is well confirmed by the general results obtained on this topic. In fact, the more the initial nitrogen content is increased by the addition of curbs, the more the mineral nitrogen will increase at the end of the composting cycle.

Thus, the contribution of olive mill wastewater to windrows rich in sludge considerably increases their nitrogen content. This nitrogen elements have the direct effect of increasing the general microbial activity and boosting the environment of the windrow.

Also, the application of olive mill wastewater has the effect of necessarily increasing the nitrogen content of the appropriate windrows. These nitrogenous elements reduce the C/N ratio and favor the stimulation of the biological activity.

As regards the co-composting of waste cuttings and straw, windrows with a large amount of straw and olive mill wastewater showed low C/N ratio values as compared to those obtained for windrows without olive mill wastewater; with respective C/N values for the windrows (A7: 0% margins, A8: 15% margins and A9: 30% margins) of 124, 63.5 and 40.3 will be recorded.

But we could notice that at the level of the finished product, a C/N ratio of 40.3 remains a rather high ratio indicating that the finished compost is not mature and an extension of the composting process is obligatorily essential because a non-mature compost might cause various important disturbances and an imbalance in the soil [19]. Thus, fertilizing substances and toxic substances such as ammonium are added to the soil in parallel. A similarity of the results would be observed in straw-based windrows where the respective values of C/N or windrows having received (0.15 and 30%) olive mill wastewater were 139; 54.38; 48.44 and 43.67.

Considering the organic components of the original waste, the addition of olive mill wastewater has increased the organic content in all the waste mass variants and this increase is proportional with the proportions of waste added for co-composting, 15% and 30%. Also, it is well known that the degradation of organic matter essentially corresponds to the mineralization of carbon and nitrogen. This result is confirmed by the fact that the smallest percentage of final residual organic matter would be recorded for windrows based on green waste. On the other hand, this percentage of residual final organic matter would be highest for straw and cuttings waste windrows, which would generally show low biodegradation.

The significant organic matter contents measured at the end of the composting cycle for sludge-based windrows could be explained by the richness of the sludge in easily metabolizable organic elements eliminated and rendered available during wastewater treatment by the complex biological flocculation-decantation-filtration phenomena.

Bacteriological analyses carried out on the various windrows at the beginning of the composting process often show great faecal bacteria contamination. Whereas an important abatement of these bacteria would be recorded at the end of the composting process. This result is certainly due to the inactivating effects of the thermophilic phase often known as the hygienization stage and exercising a significant influence on the composition of the macro and microbial flora, and consequently on the biological quality of the finished product.

These microbial inactivation effects during the thermophilic phase are very variable according to the nature of waste to be composted and the contents of the compost heap during the composting cycle. The addition of olive mill wastewater in the materials to be composted generally generates a stimulating effect of the organic biodegradation and an inhibiting and stressful effects concerning the growth of organisms presenting pathogenic or phytopathogenic characteristics [20].

The results relating to the analysis of polyphenols at the level of the windrows during the composting process (samples taken at the beginning and at the end of the composting cycle) show that the biodegradation or transformation of the polyphenols contained in the olive mill wastewater is possible by composting. Indeed, the polyphenol contents recorded at the beginning of the composting cycle are relatively higher than those obtained at the end of the composting cycle.

However, these polyphenol contents rise at the level of the windrows having received increasing quantity of 15 and 30% of the compost. At the end of the cycle, the polyphenol contents become very minimal and barely detectable (traces of the order of 0.3 mEq/g of dry matter of finished compost). The highest value is recorded at the level of the straw swath saturated with 0.55 mEq/g of finished compost dry matter.

This important result underlines the fact that the saturation of the waste windrows in the compost heap is not advisable and not recommended.

The application of a growing proportion of olive mill wastewater in the windrows does not affect the content and richness of the finished compost in mineral elements such as Zn, Cu, Ca and Cr, etc. The only exception is observed in case of the WWTPs sludge (A10, A11, and A12) with very high values of these last elements that are often proven conform to the common standards in vigor.

For the co-composting procedure treated with olive mill wastewater, we necessity assumes the improvement of the potassium and calcium contents of the finished product by assuming the great richness of the olive mill wastewater in these elements. A contrary result could be made and suggested that these nutrients have leaching with the rain in the case of processes conducted without shelter or during watering, especially in the case of potassium, which is the most soluble and therefore the most altered by leaching.

In the same manner, it is advisable to use straws (windrows: A16, A15 and A14) for composting on condition that more water or olive mill wastewater is used for watering and for avoiding low-slung water content at the start of the process that slows down the general process of composting.

Finally, industrial technology systems could therefore exploit the olive mill wastewater, which is considered as a precious raw material, very rich in organic matter and nutrients, in order to use them instead of water for watering the windrows during overall composting process.

6. Conclusion

Thus, the problem of the valorisation of the various organic residues, principally of the municipal wastes and olive mill wastewater, is currently worldwide worrying, and it is specially posed in terms of environmental preservation. The choice of the treatment technique must not be only from the point of view of monetary profitability, but above all must consider the efficiency of the treatment process. It is not possible to develop all the techniques currently being tested. Composting appeared as the main pathways for remediating this high and gigantic tonnage of wastes daily generated by modern societies. The processes of composting described above appeared very simple to be implemented, not expensive, and especially gave a finished product of good quality on average, but very consuming in space and in time. At last, I could clearly see that this kind of process might go well with an implantation especially in developing countries with a sunny and warm climate. The amendment of agricultural soil with organic matter is made very urgent by the flagrant lack of its substances on the market and the same demand for fertilisers from the increasingly demanding crops. Soil and the various bio-physico-chemical and especially biological processes of degradation and humification prevailing in the soil contribute intensively and could help in the solution of this thorny problem of accumulation of all organic residues in modern societies.

Author details

Wafa Hassen[1]*, Bilel Hassen[2], Rim Werhani[3], Yassine Hidri[4] and Abdennaceur Hassen[3]

1 Research Unit of Analysis and Process Applied to the Environmental (APAE), Higher Institute of Applied Sciences and Technology Mahdia, University of Monastir, Tunisia

2 University of Tunis El Manar, Institute for Veterinary Research of Tunisia, Tunis, Tunisia

3 Water Research and Technology Center (C.E.R.T. E), Borj-Cédria Technology Park, Soliman, Tunisia

4 Laboratory of Integrated Olive Production (LR16IO3), Sousse, Tunisia

*Address all correspondence to: hassen.wafa@gmail.com

References

[1] Hassen A., Jedidi N., K. Belguith, M. Cherif M. & Boudabous A. (2001), Microbial characterization during composting of municipal solid waste. Bioresources Technology, 80, 231-239.

[2] Mokni–Tlili S., Jaoua L., Murano F., Jedidi N. & Hassen A. (2009) Study of urban organic residues effect on culturable actinobacteria distribution in Tunisia. Waste Management & Research, 27 (3): 224-232.

[3] Mokni-Tlili S., Belguith H., Hassen A. & Gargouri A. (2011) Studies on the ecology of actinomycetes in an agricultural soil amended with organic residues: II. Assessment of enzymatic activities of Actinomycetales isolates. World Journal of Microbiology and Biotechnology, 17, 1-9.

[4] Mustin, M., 1987. Compost, Management of organic matter. F. Dubusc Eds., Paris, 954p.

[5] Ben Ammar S. (2006). Les enjeux de la caractérisation des déchets ménagers pour le choix de traitements adaptés dans les pays en développement Résultats de la caractérisation dans le grand Tunis: mise au point d'une méthode adaptée, PhD thèses, École Nationale Supérieure de Géologie de Nancy, Institut National Polytechnique de Lorraine, France, 317 pages.

[6] Fourti O., Hidri Y., Murano F., Jedidi N., Hassen A. (2008) New process assessment of co-composting of municipal solid wastes and sewage sludge in semi-arid pedo-climatic condition. Annals of Microbiology, 58 (3) 403-409.

[7] Ben Ayed L., Hassen A., Gtari M., Jedidi N., Saidi N., Jaoua L., Murano F., 2006. Évaluation de l'efficacité de trois méthodes d'extraction d'ADN de la biomasse microbienne totale du compost Déchets - Revue Francophone D'écologie Industrielle -Trimestriel - N° 44 - Parution Décembre 2006 -

[8] De Bertoldi M., Sequi P., Lemmes B., Papi T. (1996) The Science of Composting Springer; 1996th edition (February 29, 1996) ISBN-10: 0751403830, ISBN-13: 978-0751403831, 1494 p.

[9] Jammes, C. 2007. Co-valorisation of fatty effluents and lignocellulosic residues: mechanical dehydration and composting, PhD thesis, Doctoral School Sciences-Technology-Health, University of Limoges, France, 193 pages.

[10] Saidi N., Cherif M., Jedidi N., Mahrouk M., Murano F., Hassen A. & Boudabous A. (2008) Evolution of biochemical parameters during composting of various waste compost. American Journal of Environmental Sciences, 4 (4): 333-341.

[11] Canuto, RA. (2012). Dehydrogenases || Dehydrogenase Activity in the Soil Environment. IntechOpen, 10.5772/2903(Chapter 8), DOI:10.5772/48294.

[12] Jedidi N., Hassen A., Van Cleemput O. & M'hiri A. (2004) Microbial biomass in a soil amended with different organic wastes. Waste Management & Research, 22. 2:93–99.

[13] Bouzaiane O., Cherif H., Ayari F., Jedidi N. & Hassen A. (2007a) Municipal solid waste compost dose effects on soil microbial biomass determined by chloroform fumigation-extraction and DNA methods. Annals of Microbiology. Vol. 57, N° 4 December 2007.

[14] Bouzaiane O., Cherif H., Saidi N., Jedidi N. & Hassen A. (2007b). Effect of municipal solid waste compost

application on the microbial biomass of cultivated and non-cultivated soil in a semi-arid zone. Waste Management & Research. Vol. 25, N° 4, 334-342.

[15] Khan KS., Mack R., Castillo X., Kaiser M., Joergensen RG. (2016). Microbial biomass, fungal and bacterial residues, and their relationships to the soil organic matter C/N/P/S ratios. Geoderma, 271(), 115-123. DOI: 10.1016/j.geoderma.

[16] Cherif H., Ouzari H., Marzorati M., Brusetti L., Jedidi N., Hassen A. & Daffonchio D. (2008). Bacteria community diversity assessment in municipal solid waste compost amended soil using DGGE and ARISA fingerprinting methods World Journal of Microbiology and Biotechnology, 24 (7), pp. 1159-1167. DOI 10.1007/s11274-007-9588-z.

[17] Hidri Y., Bouziri L., Maron P.A., Anane M., Jedidi N., Hassen A. & Ranjard L. (2009). Soil DNA evidence for altered microbial diversity after long-term application of municipal wastewater. Agronomy for Sustainable Development, 30 (2), 423-432.

[18] Laconi S.; Molle G.; Cabiddu A.; Pompei R. (2007). Bioremediation of olive oil mill wastewater and production of microbial biomass. Biodegradation (2007) 18:559-566. DOI 10.1007/s10532-006-9087-1

[19] Fourti O., Jedidi N. & Hassen A. (2011). Humic substances change during the co-composting process of municipal. Comparison of methods for evaluating stability and maturity of co-composting of municipal solid wastes and sewage sludge in semi-arid pedo-climatic condition. Natural Science, 3(2), pp. 124-135, DOI:10.4236/ns.2011.32018.

[20] Hechmi S., Hamdi H., Mokni-Tlili S., Ghorbel M., Khelil M. N., Zoghlami I. R., Benzarti S., Jellali S., Hassen A. & Jedidi N. (2020) Impact of urban sewage sludge on soil physico-chemical properties and phytotoxicity as influenced by soil texture and reuse conditions DOI: 10.1002/jeq2.20093 Journal of Environmental Quality © 2020 American Society of Agronomy, Crop Science Society of America, and Soil Science Society of America

Chapter 4

Composting and Organic Waste Recycling a Better Option for Food Safety and Food Security

Alabi Olusoji David

Abstract

Composting is a process of degrading organic waste to form a stable material through a control process called aerobics with a biodegradation conversion process that allows the colonization of beneficial microorganisms. Careful handling of inorganic waste was significant to reducing the cost of compounding livestock feed and minimizing waste disposal in the community. Innovative technologies can inactivate pathogens in organic waste. Compost has been found to be effective in stimulating plant growth and suppressing diseases and pathogens. Safety standards remains essential for producing food rich in proteins. The use of compost can sufficiently address the challenges of malnutrition and poverty worldwide. Composting through invertebrates was also found to be significant. This is the basis upon which life exists because of the continued recycling of waste. Food security and safety goes hand in hand with the use of compost. None of these parameters must be overlooked if food production is required to meet the needs of the continuously growing population in the future.

Keywords: Compost, food safety, food security

1. Introduction

Food availability in many regions of the world has become inefficient. This has resolved into malnutrition and inadequate access among small communities in different parts of the world [1]. In this regard, innovative ideas are required to convert organic waste into compost and animal feed, which is pathogen free [1, 2]. In other words, compost is an organic substance that prioritizes food safety and security [2]. It is a source of energy that is required to provide nutrients to sustain biodiversity [2]. The inorganic components of a compost include (N, P, K, and S), secondary nutrients (Ca and Mg), and micro nutrients (B, Cl, Cu, Fe, Mn, Mo, and Zn) [2, 3]. It plays an important role in building resilience farming systems [3, 4]. It is an organic source and nutrient source in greenhouse horticulture [5]. It is a growth medium for nursery plants. It decreases the total cost associated with fertilizer and pesticides and improves nutrient retention within vegetative systems by reducing erosion and leaching [5]. Livestock manure also contains large numbers of pathogenic microorganisms [6, 7]. These organisms can pose risks to human health. These pathogens are easily transmitted from animals to humans through diverse means, such as feed processing from waste, direct contact and others. Livestock manure is used for composting [6, 7]. Some farmers applied manure directly to the farm without pretreatment [6, 7]. Livestock manure contains toxic heavy metals that may affect plant growth and reduce the efficiency of agricultural land [7, 8]. Excessive application can impair nutrient elements and are often leached beneath the reach. These toxic heavy metals not only affect plants but also animals through grazing on fertilized pasture land via ingestion [6–8]. The presence of these heavy metals (**Table 1**) in the body tissue is detrimental to human health [6–8]. These attributes of livestock manure, however call for safety and security [6–8]. The use of modern technology is required to produce a compost that is more reliable and resilient at suppressing the global distribution of soil-borne pathogens that threaten food security [5–8]. According to research findings, effective control of Rhizoctoria can be prepared from lignocellulosic substrates such as tree bark, which colonize Trichoderma spp. [9]. Compost products based on poultry are hosted by *Rhizoctonia spp. solari* and *Escherichia. coli* [9]. Composting is a biological process through which microorganisms convert organic materials into useful end products, which may be used as soil conditioners and/or organic fertilizers [9, 10]. It is a thermophilic process that stimulates the action of microorganisms to digest organic waste under aerobic conditions to produce an end product that is stable and free of pathogens. Aerated compost teas (ACT) are products in which the compost-water extract is actively aerated during the fermentation process [8–10]. Non-aerated compost teas (NCT) are products in which the compost-water extract is not aerated or receives minimal aeration only at the initial mixing stage of the fermentation process [9, 10]. Compost performs the functions of fertility and carbon sequestration in thermophilic phases [9, 10]. This allows compost to mature and be cured. One of the biggest initiatives is to develop sustainable solutions to end hunger and achieve food security through small-scale farmers and sustainable food production systems [9–11]. Research

Trace metal	Content value	Mean value
Range Mean		
Arsenic (As)	23–74	52
Boron (B)	108–240	174
Beryllium (Be)	9–39	18
Cadmium (Cd)	2.5 14.8	7.5
Cobalt (Co)	6–84	48
Chromium (Cr)	100–280	170
Copper (Cu)	280–1100	610
Mercury (Hg)	15–25	21
Manganese (Mn)	385–1600	800
Molybdenum (Mo)	20–35	25
Nickel (Ni)	90–180	140
Lead (Pb)	385–4100	1630
Selenium (Se)	3.5–6.9	4.8
Titanium (Ti)	2200	2200
Vanadium (V)	38–310	170
Zinc (Zn)	465–2250	1350

Table 1.
Trace elements and heavy metal available in untreated livestock waste source: [7].

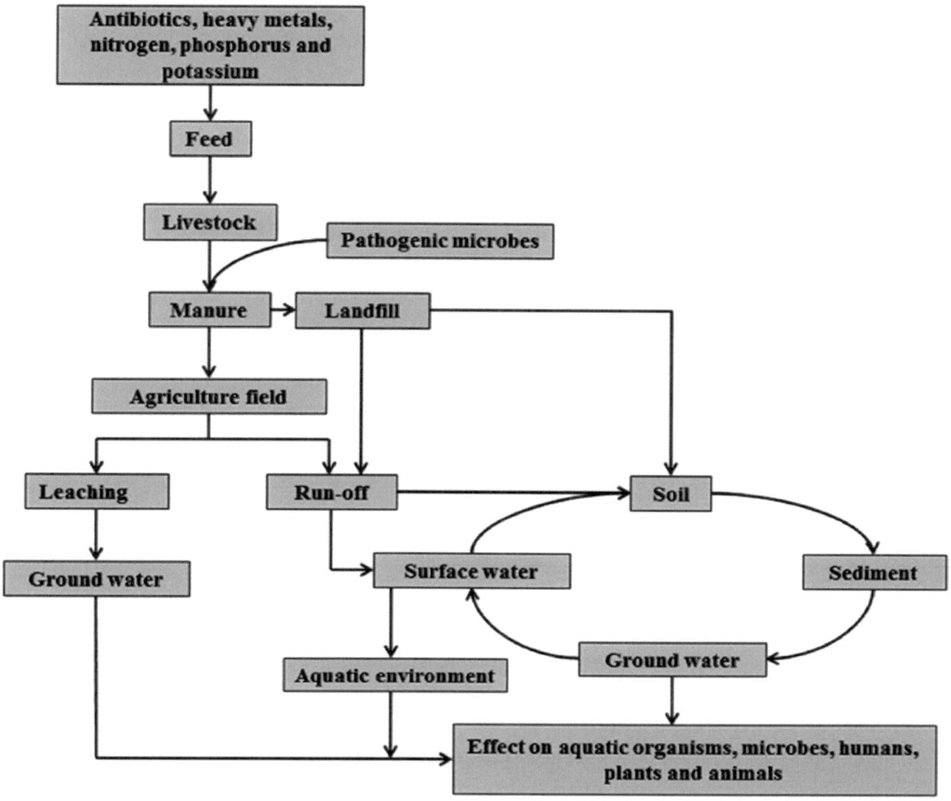

Figure 1.
Influence of untreated livestock waste on the ecosystem sourced: [11].

has focused on identifying the most sustainable and safe strategies for food waste [9–11]. This is a priority for turning waste into a resource by valorizing them into the agri-food supply chain [9–11]. **Figure 1** above, however describes the influence of untreated livestock waste on the ecosystem.

2. Organic waste as a supplement in animal feed

According to the United Nation findings in 2019, the world population was estimated to be 7.7 billion and was projected to be 9.7 billion in 2050 [2, 3]. In the recent times, hunger and food insecurity has become a global problem [2, 3]. Undernourishment or micronutrient deficiency was aggravated due to increasing population. Likewise, health challenges e.g. obesity may become an issue by 2025 [2, 3]. This is because human nutrition increases with increasing consumption of sugar, oils and fat compared to proteins [4]. Looking at the future, the demand for animal product such as meat may become unsustainable by 2050 [4]. In order to make provision for this, meat production is expected to double the amount of 455 million tons by 2050 [2, 3]. The production of poultry is increasing at alarming rate compared to livestock production [12]. The consumption of chicken meat and eggs has become a major source of proteins across the globe. Unfortunately, the food production may not be proportionate to human consumption due to finite natural resources [12, 13]. Urban development and industrial revolution has drastically influence the availability of land for food production. It was also estimated that

1.3 billion tons of food are produced globally and are wasted during production, postharvest, and processing [12, 13]. A huge portion of organic waste include edible and inedible food. Organic waste generated at the latter part of food supply is comparatively higher than at the early stage of the chain. It was also projected that increasing demand for animal product may result in high demand for feed particularly the coarse grains such as maize and protein meal by 2025 [12–14]. The conventional poultry and swine diet depend on maize and soyabean as energy and proteins sources [12–14]. As well, these same, staple food is largely consumed in many part of the world particularly in the developing countries [13, 14]. As a result, a notable increase in the cost of staple food for animal feed drives the cost of production. However, the likely solution for this could be organic waste from cereal grains and plant proteins sources used in animal nutrition. However, production cost is reduced with introduction of food waste due to its low cost compared to conventional feeds [13–15]. Food organic waste from poultry, cattle, etc. is mixed with livestock feed. This reduces the amount of waste disposal in the community [13–15]. In other words, waste retains nutritive value, which may reduce the cost of feeding. However, the waste, the type and quantity of other foodstuffs, the way the excrement is treated and the species to which the treated excrement is fed is related to its efficiency [14, 15].

3. The shelf life of processed organic waste and relevance

The efficiency of organic waste largely depend on its safety at acquiring security. This is required particularly when composing organic waste with the feed [16]. This prolonging the shell life of processed organic waste [17, 18]. The treatment of organic waste can be obtained by cooking, extrusion, pellatizing, dehydration, ensiling and probiotics [17, 18]. The shelf life of the organic waste are better prolonged with extrusion, pellatizing, and dehydration than cooking and ensiling [19]. The dehydration treatment is helpful when considering the treatment of resultant food waste in swine [19, 20]. The moisture content of organic waste varies between 50% and 85%. This may threatened the use of organic waste in animal feed but the application of adequate treatment reduced the moisture content to produce a safe organic waste [19, 20]. Adequate heat could prevent organic waste susceptibility to oxygen. This helps to prolong shelf life of waste during storage and as well prevent lipid peroxidation [20, 21]. Solar energy was identified to be economical and environmental viable at transforming organic waste into dried waste. Solar energy was used at pasteurizing and drying of organic waste. This process produce low carbon emissions and energy efficient [20, 21]. Likewise in a findings, it was also revealed that 92.74% dry matter was obtained from the use of solar energy. The crude proteins, EE and crude fiber content were 25.68%, 21.57% and 0.75% respectively [21, 22].

Furthermore, the quality and quantity of dietary proteins are of importance in poultry and swine feed [21, 22]. 15% - 23% of crude proteins was found in organic waste on a dry basis [22, 23]. The C.P obtained in organic waste in an experiment on a dry basis was estimated to be 18.9%. This was significantly higher than the conventional feed [22, 23]. The amino acid content of organic waste was higher compared to that of conventional feed [22, 23]. It was also revealed that the lysine, methionine, threonine and tryptophan content of organic waste was higher compare to the conventional feed while a similar result was obtained comparing the two diets [24]. Phosphorus and calcium are important element relevant for cell division and tissue formation. The availability of phosphorus and calcium was studied in an investigation [24]. It was found that organic waste contain 0.004% to 0.46%

calcium on a dry basis [24, 25]. This has a similarity with result obtained with soyabean meal 0.27% - 0.47% with a C P of 47.2%. This was higher than result obtained from corn (0.01–0.03%) [24, 25]. The breeding of broiler was experimented using different concentration of organic waste. Result revealed that the dried leftover food produced 20.02% C P and 9.99% EE [25]. The concentration of DHA in meat from broiler group fed dried leftover was not significantly higher than the control group. It was also recorded that the arachidonic acid content in broiler meat was not different among the treatment. However, the leftover treatment produced a better result than the control [25]. The salt content in organic waste compared with conventional feed made from corn in an experiment. It was reported that salt content of organic waste was higher compare to the conventional feed [25, 26]. The salt content of organic waste was recorded to be between 2.0% to 2.5% and approximately 3.28% on a dry basis. The present of salt in animal diet was very helpful. It enhances palatability [26–28]. It prevent the growth mold. It regulates growth and performance of the animals [28, 29]. Adequate salt level provide a soft texture of pork. It was recorded in a study that the salt content of organic waste was higher compare to convention feed composed from corn and soyabeans [28, 29]. It was also found that the performance of poultry and pig was strictly due to high salt content. Organic waste can be used as a substitute for salt if properly incorporated into diet because high salt level may result into poisoning [29]. It may also be subjected to rancidity [29]. Animal feed contribute to alleviation of food security through utilization of organic waste. Some bioactive compound and nutrients are found helpful to animal feed [29]. The use of organic waste in poultry and pig was found to be promising.

4. Food waste and pathogens

Moreover, waste generated by animal compounds is a more serious problem than waste voided by humans. Human waste is treated before it is released into waterways, but animal waste is not or poorly processed [30]. It has been reported that disease organisms may be transmitted from animal to animal through the use of organic manure in feed processing. The transfer of these diseases may constitute a problem to animal and public health [31]. Chicken waste is well known for Salmonella and Campylobacter spp. [32]. Cattle are known to host a gram of cattle manure with up to 10 million Salmonella organisms [31, 32]. Several methods have been used by researchers to inactivate pathogens before they are introduced into animal feed. The use of the stack method was found to be effective by some researchers [31, 32]. Some Salmonella species require an approximate temperature of 63°C (145°F), whereas *E. coli* requires heat of approximately 68°C (153°F) [31, 32]. Fermentation was also used to breakdown the bacteria in manure [33, 34]. This was an ensilage process. It involves the ensiling of waste with various grasses stored in airtight silo. In this situation, higher temperatures is ideal for killing infectious organisms [34]. Under these conditions, poultry litter buffers acids that destroy bacteria [35, 36]. In other words, some bacteria and coliforms were not completely controlled by ensilage. In a different finding, it was also recorded that stacking and fermentation is not as effective as the use of heating and drying method [35–37]. This involves subjecting manure or litter to the hot air blast from the oven. The temperature here is approximately 150°C (300°F) or higher, depending on the method [36, 37]. Campylobacter jejuni, *E. coli* Samonella, germisia and Listeria were found to survive prolonged anaerobic digestion [36, 37].

The use of organic manure in feed processing allows the transfer of disease-causing bacteria to spread resilience strains that may not be associated with organic waste. Most of these diseases inactivate at different temperatures [35–37].

Disease	Organism	Lethal conditions (moist heat)
Anthrax	*Bacillus anthracis*	Over 100°C
Brucellosis or Contagious abortion	Brucella species	10 mins - 60°C
Cholera	Vibrio cholera	15 mins - 55 °C
Diphtheria Diphtheria	Corynebacterium	10 mins - 58°C
Dysentry	Shigella species	60 mins - 55 °C
Food poisoning	Salmonella species	20 mins - 60 °C
Leptospirosis (Weil's disease)	Leptospira species	10 mins - 50 °C
Plague	Pasteurella pestis	5 mins - 55 °C
Staphylococcal infections	Staphylococci	30 mins - 60 °C
Streptococcal infections	Streptococci	30 mins - 55°C
Tuberculosis	Mycobacterium	20 mins - 60 °C
Typhoid fever	Salmonella typhi	20 mins - 60 °C
Vibriosis Intestinal worms	Vibrio fetus	5 mins - 56°C
Round worm	Ascaris lumbricoides	60 mins - 55 °C
Tape worm	Taenia saginata	A few minutes - 55 °C
Hookworm	Ancylostoma duodenale and *Necator americanus*	1 min - 55 °C

Table 2.
Lethal- time conditions for some common pathogen sourced: [2].

In addition, European safety rules for food should be followed by a high level of protection of human health, and the production of edible terrestrial invertebrates as food should be safe and wholesome and reared and marketed as food [37–42]. The lethal time conditions for some common pathogens are shown in the **Table 2**.

5. Invertebrate function as food and compost

The population of certain invertebrate are excellent source of biologically valuable protein, micronutrients, minerals, and vitamins in the human diet [43]. This species contains high protein content in the range of 54.6–71% dry matter and is rich in amino acids considered essential for humans. It also rich in fat, with content ranging from 7.3–10% of dry matter [43]. Certain invertebrate such as earthworm are grown in organic waste in a confined area for consumption purpose and vermicomposting [43, 44]. European safety rules for food must be followed with high level of protection for human health. The production of edible terrestrial invertebrates as food must be safe and wholesome for consumption and for market purpose [43, 44]. This invertebrate meal such as earthworm, millipede therefore has interesting nutritional proprieties but in order to be commercialized as a product for human consumption and/or animal feed, it must be safe for final consumption [44]. Safety standards must be ascertained to ensure disease organism are cured from these invertebrates before they are introduced for consumption [45, 46]. The finished food product must undergo assessment and as well ensure its hygienic and undesirable substances. To achieve the safety on food product, it is necessary to evaluate the microbiologic profile, testing for microorganism indicators of process hygiene and verifying compliance with food safety parameters, chemical profile, the presence of pesticide residue and toxic elements [45, 46]. Similarly, the growth of earthworm provides sufficient vermicompost as an organic fertilizer that allows a reduction in the use of mineral fertilizers in other production systems [45, 46]. For food purposes, earthworms for instances were subjected to a deep washing and vacuum freeze-drying procedure, a useful dehydration method is used to preserve nutritional quality [47–50]. Then, the earthworm meal are defatted to obtain a meal rich in protein and other nutrients and to decrease perishability. Food are free of pesticides or mycotoxin. The meal possesses safety traits in relation to possible public health risks [51–54].

6. Vermicompost sustainable means for food security

Invertebrate communities are made up of detrivores, microbivores, predators that are functional at engineering ecosystem [55]. The physical and chemical composition of the soil may be related to the present of certain invertebrate [56]. The invertebrate communities perform the function of imputing organic matter into the soil through frequent mulching, compost, biosolid into soil habitat [55, 56]. It involves the breaking down of tree leaf, litter to grass clipping and root exudate [55, 56]. The availability and the abundant of these invertebrate determines the efficiency of the ecosystem. The invertebrate apart from earthworm that are found effective at composting organic waste this include dipteran larva, fruit fly (Drosophila spp.) and the P (Dipteran) [55, 57]. Dipteran larva composes waste at a mesophilic temperature not higher than 40°C but fruit fly (Drosiphila spp) has proven to be efficient at a thermophilic temperature greater than 40°C. The P (Dipteran) was found to have cured waste at a temperature less than 40°C [56–58]. Earthworm had been the most common invertebrate well known at composting. Vemicompost has been used in bioconversion of fruit and vegetable waste releasing organic soluble for plant growth, which is sustainable, cost-effective, and ecological approach that can contributes to the reduction of food waste [56–58]. Fruit and vegetable waste as a substrate for the growth of earthworms produces a valuable product that can influence agricultural land and crops [57, 58]. In nowadays, scientist consider use of terrestrial invertebrates such as earthworms used as an alternative source valid at restoring land productivity, especially by reintroducing fruit and vegetable waste into the food supply chain, hence turning waste into a resource [58]. Pressmud is a byproduct of bagasse. Pressmud contains about 25–30% organic matter. The major plant nutrient constituent of pressmud include N, P, K, Ca, Mg & S and the minor elements like Fe, Zn, Mn, Cu, B & Mo. Pressmud can be used in vermicomposting [58, 59]. The population of earthworm called Eisenia foetida are very effective and excellent source of biologically micro and macronutrients [58, 59]. Recycling biodegradable waste into compost and compost tea is being promoted as a viable option for treating waste material [60]. As well as the use of compost as a bulk fertilizer and soil ameliorant. A considerable evidence also shows that compost and liquid preparations such as compost tea can suppress soil-borne diseases [60]. However, the effect of compost and compost tea on soil-borne diseases varies greatly depending on the properties of the compost as affected by compost formulation, the composting and compost tea brewing process, and the environmental conditions in which the material is used [60, 61]. According to [61], the efficacy of biological control as trichoderma harmatum increases as compost mature and the ratio of cellulose lignin than hay or straw based on composts. Mature compost with wood chips or bark and anaerobic digestion suppresses Rhizoctonia [60, 61]. Compost with wood as an ingredient passing through the gut of earthworm provides a healthy micro biome that provide ecosystem functions including promotion of plant growth. The microbes however, mediates and prevent limiting organisms to releasing plant nutrient. This indicate that beneficial microbe compete with antagonizing pathogen [60, 61]. Likewise, these microbial species produce antibiotics to combat pathogens, manufacture plant growth hormone or induce systematic plant defense that promote plant growth. However, a good composition and curing method deliver a national consortium of microbes for biological control of diseases [60, 62].

7. Composts suppress diseases organisms in the root

Disease suppressiveness vary differently with respect to pathogens for instance some soil with compost were found suppressive to Fusarium wilt [62]. The suppression of diseases are specific to composition and the various components blended together during composting. A compost that is steam sterilized are found efficient and stable [62, 63]. Pythium and root knot nematode was found suppressed with the application of organic manure in the growth of tomato, sweet peppers and ornamental [64, 65]. Soil borne diseases was also found suppressed with the use of compost during plant growth [66, 67]. Mature compost suppresses diseases with the addition of organic amendment and biological control agent [66, 67]. The use of compost increases the suppression of diseases. Compost application prevent quick re-infestation of soil by pathogen. Diseases such as bacterial cancker and wilt of tomato are caused by *Clavibacter michiganensis* and *Ralstonia solanacearum* [67]. It was also recorded that compost was effective at reducing botrytis cinerea on cucumber and melon bacterial leaf spot of *Xanthomonas campestris* on tomato [68]. The control of these pathogens minimized the invasion of diseases. However, the security of crops using mature compost during vegetative growth and fruiting is significant to increasing farm output. As a result, the palatability of crops are enhanced. The appearance of crop are appealing to the eyes. More so, the volume of harvest in the market for final consumer is at maximum [67, 68]. Compost can be used as alternative to synthetic fertilizer which has been found to be detrimental to ecosystem through surface runoff [69, 70]. Most farmers in attempt to increase yield, applies synthetic fertilizer in excess. Excessive use of synthetic fertilizer increases yield at a diminishing rate because continuous addition of synthetic fertilizer become toxic to beneficial organisms and as well reduce yield [70]. Synthetic fertilizer infestation pests and disease organisms through competition from weed [71, 72]. It was also recorded in a findings that excessive use of synthetic fertilizer can accelerate early rot of food crop and vegetables before and after harvest [71, 72]. The palatability of food crop are negatively affected with the use of synthetic fertilizer. However, mature compost are naturally inclined. They produce beneficial organisms that fight pathogens. Mature compost releases nutrient gradually released into the soil. It increases soil moisture and aeration. Compost application to vegetables and food crop can be helpful at achieving food security if properly managed [73].

8. Conclusions

Invertebrate plays a significant in carbon cycle and green community development. It is a good source of protein. Little or no attention are given to these organisms. They provide the required energy needed to sustain biodiversity through composting. However, compost has been identified as organic component that contains requires micro and macro nutrients that is capable of restoring nutrient into the soil. In this case, plant and animal waste are of significant interest. Waste are subjected to either aerated or non-aerated temperature to produce a pathogen free compost. The end product contains some compounds that could stimulate development of plant and animal. Pretreatment are required at an appropriate temperature essentially to inactivate disease organisms and pathogens. Compost has been used as a supplement for animal feed, it reduces the cost of production. Likewise, disposal of organic waste in waterways or occupying large land mass in the community. Compost has been used to boost germination and growth of plant. It suppresses wide range of disease varying from one plant to the other. Likewise, safety standard must be ensured and storage of food to keep the public free of zoonotic diseases. In a nutshell, if compost are properly managed, it is capable of increasing food production. Wiping away malnutrition and poor access to food across the globe.

Author details

Alabi Olusoji David
Mangosuthu University of Technology, Durban, South Africa

*Address all correspondence to: oludavidsj@gmail.com

References

[1] Papargyropoulou, E. Lozano, R. Steinberger, J. K. Wright, N. B. Ujang, Z (2014). The food waste hierarchy as a framework for the management of food surplus and food waste. Journal of Cleaner Production.

[2] FAO, (1995). Soil management compost production and use in tropical and subtropical environmental. Soils Bulletin No. 56(6), 707- 715.

[3] Jurgilevich, A. Birge, T. Kentala-Lehtonen, J. Korhonen-Kurki, K. Pietikäinen, J. Saikku, L. Schösler, H (2016). Transition towards Circular Economy in the Food System. Sustainability. 8, 69.

[4] Parry, A., P. Bleazard and K. Okawa (2015), "Preventing Food Waste: Case Studies of Japan and the United Kingdom", OECD. Food, Agriculture and Fisheries Papers, No. 76, OECD.

[5] Zhous, W. Baosheng, Qlu. (2005). Effect of cadmium hyperaccumulation on physiological characteristics of Sedum alfredii Hance (Crassulaceae). Plant Science 169 (4), 737 745.

[6] Ramasamy Rajesh Kumar, Jae Young Cho (2013). Application and Environmental Risks of Livestock Manure. Article *in* Journal of the Korean Society for Applied Biological Chemistry. (30) 1,124.

[7] FAO (2020). Animal Nutrition in FAO and Sustainable Development Goals; FAO: Rome, Italy, 2017. Foods. 9, 291 17 of 18.

[8] Zu Ermgassen, E.K.; Phalan, B.; Green, R.E.; Balmford, A (2016). Reducing the land use of EU pork production: Where there's will there's a way. Food Policy. 58, 35-48.

[9] Dargatz, D. (1995). Shedding of Eschericia coli 0157:H7 by feedlot cattle. In: Hurd WS, Eav L, editors. APHIS report of accomplishments in animal production food safety FY 1995. Washington: USDA, Animal and Plant Health Inspection Service. 1, 7-1-1.7-2.

[10] Vielma, R.A. Carrero, P. Rondon, C. Medina, A.L. (2007). Comparison of the content of minerals and trace-elements in flour of earthworm (Eisenia foetida) using two drying procedures. Saber Rev. Multidisciple Construction Investigation. University Oriental 2007, 19, 83-89.

[11] Ramasamy Rajesh Kumar, Jae Young Cho (2013). Application and Environmental Risks of Livestock Manure. Article *in* Journal of the Korean Society for Applied Biological Chemistry. (30) 1,124.

[12] Alexandratos, N. Bruinsma, J. (2012) World Agriculture Towards 2030/2050: The 2012 Revision; Food and Agriculture Organization of the United Nations, Agricultural Development Economics Division (ESA): Rome, Italy.

[13] Stenmarck, A. Jensen, C. Quested, T. Moates, G (2016). Estimates of European Food Waste Levels; Swedish Environmental Research Institute: Stockholm, Sweden.

[14] Truong, L.; Morash, D.; Liu, Y.; King, A. Food waste in animal feed with a focus on use for broilers. International Journal Recycling Organic Waste Agric. 8, 417-429.

[15] Garcia, A.J. Esteban, M.B. Marquez, M.C. Ramos, P. (2020). Biodegradable municipal solid waste: Characterization and potential use as animal feedstuffs. Waste Management 2005, 25, 780-787. Foods. 9, 291 16 of 18.

[16] Santos, D.I. Saraiva, J.M.A. Vicente, A.A. Moldão-Martins, M (2019). 2—Methods for determining

bioavailability and bioaccessibility of bioactive compounds and nutrients. In Innovative Thermal and Non-Thermal Processing, Bioaccessibility and Bioavailability of Nutrients and Bioactive Compounds; Barba, F.J., Saraiva, J.M.A., Cravotto, G., Lorenzo, J.M., Eds.;Woodhead Publishing: Sawston, UK. pp. 23-54.

[17] Fung, L. Urriola, P.E. Baker, L. Shurson, G.C (2018). Estimated energy and nutrient composition of different sources of food waste and their potential for use in sustainable swine feeding programs. Translation. Animal Science. 3, 359-368.

[18] Myer, R.O. Brendemuhl, J.H. Johnson, D.D (2000). Dehydrated Restaurant FoodWaste as Swine Feed. In FoodWaste to Animal Feed;Westendorf, M.L., Ed.; Iowa State University Press: Iowa City, IA, USA. pp. 113-144.

[19] Rojas, O.J. Stein, H.H (2017). Processing of ingredients and diets and elects on nutritional value for pigs. Journal Animal Science Biotechnology. 8, 48.

[20] El-Ahraf, A. Willis, W.V. Saleh, R. (1984). Animal waste problems and management techniques: a review and bibliography. Dairy Food Sanitation. 4,168-73.

[21] Chae, B.J. Choi, S.C. Kim, Y.G. Kim, C.H. Sohn, K.S (2000). Elects of feeding dried food waste on growth and nutrient digestibility in growing-finishing pigs. Asian Australas. Journal. Animal Science. 13, 1304-1308.

[22] Choe, J. Moyo, K.M. Park, K. Jeong, J. Kim, H. Ryu, Y. Kim, J. Kim, J.M. Lee, S. Go, G.W (2017). Meat Quality Traits of Pigs Finished on Food Waste. Korean Journal Food Science Animal. 37, 690-697.

[23] Donati, E (2018). Advances in dairy products. In Advances in Dairy Products; Contó, F., Del Nobile, M.A., Faccia, M., Zambrini, A.V., Conte, A., Eds.; JohnWiley & Sons Ltd.: Hoboken, NJ, USA.

[24] Beski, S.S.M. Swick, R.A. Iji, P.A (2015). Specialized protein products in broiler chicken nutrition: A review. Animal Nutrition. 1, 47-53.

[25] Asar, E. Raheem, H. Daoud, J (2018). Using dried leftover food as non-traditional feed for Muscovy duck diet. Assiut. Vetinary Medicine Journal, 64, 107-114.

[26] Castrica, M. Tedesco, E.A.D. Panseri, S. Ferrazzi, G. Ventura, V. Frisio, G.D. Balzaretti, M.C (2018). Pet food as the most concrete strategy for using food waste as feedstuff within the European Context: A Feasibility Study. Sustainability. 10, 2035.

[27] Martín Ortega, A.M. Segura Campos, M.R (2019). Chapter 13—Bioactive Compounds as Therapeutic Alternatives. In Bioactive Compounds; Campos, M.R.S., Ed.; Woodhead Publishing: Sawston, UK. pp. 247-264.

[28] Cheraghi Saray, S. Hosseinkhani, A. Janmohammadi, H. Zare, P. Daghighkia, H (2014). Thermal and probiotic treatment elects on restaurant waste for incorporation into poultry diet. International Journal Recycle. Organic Waste Agriculture. 3, 7.

[29] Alexandros, G. Elisavet, G. Athanasios, C. Pappas, G. P. Fenia, G. Thrassyvoulos, M. Eleni, T. K. F & George, Z (2020). Bioactive Compounds in Food Waste: A Review on the Transformation of Food Waste to Animal Feed. Foods. 9, 291.

[30] McCaskey, T.A. Martin, J, B. (1988). Evaluation of a process for improved quality and microbial safety of broiler litter. Biological Wastes. 25,209-18.

[31] Mourad, M. (2016). Recycling, recovering and preventing "food waste":

Competing solutions for food systems sustainability in the United States and France. Journal Clean Production. 2016, 126, 461-477.

[32] Moriarty, K. (2014). Feasibility Study of Anaerobic Digestion of Food Waste in St. Bernard, Louisiana; tech. Office of Sound Material-Cycle Society. History and Current State of Waste Management in Japan.

[33] Salemdeeb, R. Zu Ermgassen, E.K. Kim, M.H. Balmford, A. Al-Tabbaa, A (2017). Environmental and health impacts of using food waste as animal feed: A comparative analysis of food waste management options. Journal Clean Production. 140, 871-880.

[34] D'Aoust, J.Y. (1991). Pathogenicity of foodborne Salmonella. International Journal Food Medicine. 12, 17-40.

[35] Bryan, F.L. Doyle, M.P. (1995). Health risks and consequences of salmonella and Campylobacter jejuni in raw poultry. Journal Food Protection. 58,326-44.

[36] Caswell, L.F. Fontenot, J.P. Webb, K.E. (1978). Fermentation and utilization of broiler litter ensiled at different moisture levels. Journal Animal Science. 46, 547-61.

[37] Dou, Z. Toth, J.D. Westendorf, M.L (2018). Food waste for livestock feeding: Feasibility, safety, and sustainability implications. Global Food Security.17, 154-161.

[38] Boda, K. (1990). Nonconventional feedstuffs in the nutrition of farm animals. New York: Elsevier. 139-252.

[39] Food Wastage Footprint Full-Cost Accounting: Final Report; Food and Agriculture Organization of the United Nations, 2014.

[40] Fontenot, J.P. Webb, K.E. (1975). Health aspects of recycling animal wastes by feeding. Journal Animal Science. 40, 1267-77.

[41] Strauch, D. Hygiene of animal waste management. In: Strauch D, editor. Animal production and environmental health. New York: Elsevier, 1987.

[42] Girotto, F. Alibardi, L. Cossu, R (2015). Food waste generation and industrial uses: A review. Waste Management. 45, 32-41.

[43] Taylor, J.C. Geyer, R.E. (1979). Regulatory considerations in the use of animal waste as feed ingredients. Journal Animal Science.48, 218-22.

[44] Global Initiative on Food Loss and Waste Reduction; Food and Agricultural Organization of the United Nations, 2014.

[45] Leib, E.B. Balkus, O. Rice, C. Maley, M. Taneja, R. Cheng, R. Civita, N. Alvoid, T. Leftovers for Livestock (2016): A Legal Guide for Using Excess Food as Animal Feed; The Harvard Food Law and Policy Clinic and the Food Recovery Project at the University of Arkansas School of Law: Fayetteville, AR, USA.

[46] Kearney, T.E. Larkin, M.J. Levett, P.N. (1993). The effect of slurry storage and anaerobic digestion on survival of pathogenic bacteria. Journal Applied Bacteriol. 74, 86-93.

[47] Sun, Z. Jiang, H. (2017). Nutritive Evaluation of Earthworms as Human Food. Future Food 2017, 37.

[48] Lim, S.L. Lee, L.H. Wu, T.Y. (2016). Sustainability of using composting and vermicomposting technologies for organic solid waste biotransformation: Recent overview, greenhouse gases emissions and economic analysis. Journal Clean. Production. 111, 262-278. [CrossRef]

[49] Tedesco, D. E. A, Castrica, M, Tava, A. Panseri, S. & Balzaretti, C. M.

(2020). From a food safety prospective: The role of earthworms as food and feed in assuring food security and in valuing food waste. Insects, 11(5), 293.

[50] Pogorzelska-Nowicka, E. Atanasov, G.A. Horba´ nczuk, J.Wierzbicka, A (2018). Bioactive Compounds in Functional Meat Products. Molecules. 23, 307.

[51] Singh, M. Bhuvnesh, V. Shwet, K. Wakchaure. (2011). Mushrooms: cultivation, marketing and consumption. Cab direct.

[52] Seo, Y (2013). Current MSW Management and Waste-to-Energy Status in the Republic of Korea. M.S. Thesis, Columbia University, New York, NY.

[53] Herriott, D.E. Ebel, E. Carpenter, L. Hancock, D.D. Rice, D. Besser, T.E. (1995). Sources of verotoxic Escherichia coli 0157:H7 in feedlots and dairy farms in the Pacific Northwest. In: Hurd WS, Eav L, editors. APHIS report of accomplishments in animal production food safety FY 1995. Washington: USDA, Animal and Plant Health Inspection Service. 1, 1-1-1.1-12.

[54] Vergnes, A. Blouin, M. Muratet, A. Lerch, T. Z. Mendez-Millan, M. Rouelle-Castrec, M. (2017). Initial conditions during Technological implementation shape earthworms and ants diversity, Landscape, Urban Plan. 159, 32-41.

[55] Jurgilevich, A. Birge, T. Kentala-Lehtonen, J. Korhonen-Kurki, K. Pietikäinen, J. Saikku, L. Schösler, H (2016). Transition towards Circular Economy in the Food System. Sustainability. 8, 69.

[56] Bray, N. Kao-Kniffin, J. Frey, S. D. Fahey, T. J. & Wickings, K (2019). Soil macroinvertebrate presence alters microbial community composition and activity in the rhizosphere. Front Microbiology

[57] Oni, O.D. Elizabeth O. O, Mynepalli, K.C. Sridhar & Taiwo B. H (2020). Macro-invertebrate population changes during composting of organic waste at Alesinloye Market, Ibadan. International Journal Agriculture Innovation Technology and Globalization, Vol. 1, No. 3. 2020

[58] Huang, K. Xia, H. Li, F. Wei, Y. Cui, G. Fu, X. Chen, X. (2016). Optimal growth condition of earthworms and their vermicompost features during recycling of five different fresh fruit and vegetable wastes. Environ. Science Pollution Resource. 23, 13569-13575.

[59] Diaz, P.M. (2016). Consequences of Compost Press Mud as Fertilizers. DJ International Journal of Advances in Microbiology & Microbiological Research.

[60] Froetschel, M. A. Ross, C. L. Stewart, R. L. Azain, M. J. Michot, P. Rekaya, R (2014). Nutritional value of ensiled grocery food waste for cattle. *Journal of Animal Science*., 92(11), 5124-5133.

[61] Santos M, Dia´nez, F. Carretero, F. (2011). Suppressive effects of compost tea on phytopathogens. In: Dubey NK, editor. Natural Products in Plant Pest Management. Oxford (UK): CABI Publishing. p. 242-262.

[62] Nelson, E.B. Kuter, G.A. Hoitink, H.A. (1983). Effects of fungal antagonists and compost age on suppression of Rhizoctonia damping-off in container media amended with composted hardwood bark. Phytopathol. 73, 1457-1462.

[63] Neher, A. D. Marie, A. Limoges, T. R. Weicht, Patricia, D. Milner. (2020). Bacterial community dynamics distinguish poultry compost from dairy and non-amended soil planted with spinach Microorganism. 8 (10), 1601.

[64] Anne, B. Geoffroy, S. Cherel, J. Warot, G. Marie, S. Noël, C. J (2018).

Towards an operational methodology to optimize ecosystem services provided by urban soils. Landscape. Urban Plan. 176, 1-9.

[65] Los Angeles Department of Parks and Recreation (2016). Los Angeles Countywide Comprehensive Parks and Recreation Needs Assessment. Available online at: https://lacountyparkneeds.org/final-report.

[66] Szlavecz, K. Yesilonis, I. & Pouyat, R (2018). "Soil as a foundation to urban biodiversity," in Urban Biodiversity, eds A. Ossola and Journal. Niemela (Abingdon: Routledge in Association with GSE Research), 18-35.

[67] Stoffella, P.J. Graetz, D.A. (1996). Sugarcane filtercake compost influence on tomato emergence, seedling growth, and yields. In: Bertoldi MD, Sequi P, Lemmes B, Papi T, editors. The science of composting: Part 2. Glasgow (UK): Blackie Academic and Professional. p. 1351-1356.

[68] Litterick, A. Wood, M. (2009). The use of composts and compost extracts in plant disease control. In: Walters D, editor. Disease control in crops: Biological and environmentally friendly approaches. Oxford (UK): Wiley-Blackwell. p. 93-121.

[69] Gardebroek, C. Hernandez, M. A. Robles, M. (2014). Market Interdependence and Volatility Transmission among Major Crops; working paper. International Food Policy Research Institute.

[70] Singh, R.P. Embrandiri, A. Ibrahim, M.H. Esa, N. (2011). Management of biomass residues generated from palm oil mill: Vermicomposting a sustainable option. Resource Conservation. Recycle. 2011 55, 423-434.

[71] Tresch, S. Frey, D. Le Bayon, R.-C. Zanetta, A. Rasche, F. Fliessbach, A (2019). Litter decomposition driven by soil fauna, plant diversity and soil management in urban gardens. Science. Total Environment. 658, 1614-1629.

[72] Lopez Alonso marta, Marta Miranda, cristina Castillo, josquin Hernandez (2002). Toxic and essential metals in liver, kidney and muscle of pigs at slaughter in Galicia (NW Spain). Food Additives and contaminants 24(9), 943-54.

[73] Salvia, A.L. Leal, F. W. Brandli, L.L. Griebeler, J.S. (2019). Assessing research trends related to Sustainable Development Goals: Local and global issues. Journal Clean Production. 841-849.

Chapter 5

Role of Humic Substances in Formation of Safety and Quality of Poultry Meat

Inna Vladimirovna Simakova, Alexey Alekseevich Vasiliev, Konstantin Vyacheslavovich Korsakov, Lyudmila Alexandrovna Sivokhina, Vladimir Vasilievich Salautin, Lyudmila Yurievna Gulyaeva and Nikita Olegovich Dmitriev

Abstract

The purpose of this chapter was to study the influence of humic substances on the formation of the safety and quality of poultry meat. The high abilities of the natural and organic complex of humic acids "Reasil®HumicHealth" (produced in Russia, Saratov) to sorb and desorb five mycotoxins of compound feed (aflatoxin B1, ochratoxin, toxin T-2, zearalenone and fuminisin B1) were experimentally established. The hepatoprotective ability of humic acids was observed in experiments on broilers using rapeseed meal containing an increased amount of secondary plant metabolites that could cause liver damage. It was found out that the inclusion of humic acids in the amount of 1 and 1.5 g per 1 kg of feed to the main diet of broiler chickens has a more stable positive effect, both in terms of slaughter indicators (yield of semieviscerated carcasses and carcasses of complete evisceration), and in the production of the most valuable natural semi-finished products (breast, chicken legs) due to the intensive growth of muscle tissue. A clear improvement in the morpho-biochemical and immunological parameters of blood and micromorphometric characteristics of the organs of the immune system of poultry was noted with the use of different concentrations of humic acid salts. The use of the additive at the rate of 1 and 1.5 g per 1 kg of feed does not negatively affect the sensory indicators of finished culinary products.

Keywords: humic acids, broiler chickens, safety, quality

1. Introduction

Intensive poultry farming technologies have created the prerequisites for providing the population with dietary meat and high-grade protein. Poultry products have become in demand and, at the same time, they are a source of microbiological, physical and chemical risks. Regular consumption of poultry meat rearing using intensive technologies is the risk of a number of negative effects on human health, including antibiotic resistance of the body. Such factors negate the health benefits of such meat, especially when it is used in medical dietary or baby food.

In this regard, an important task of science and practice is the search for new ways to eliminate the toxic load on agricultural poultry, through the use of environmentally friendly biologically active substances in its diet, which have a stimulating effect on nonspecific factors of the body's natural resistance and increase productivity.

A promising direction in poultry feeding is the use of natural biological compounds - humic acids and their salts - humates, which, after interaction with a living cell, contribute to the output of released energy not on the elimination of adverse environmental influences, but on growth and reproduction, which leads to an increase in the competitiveness of an organism.

The purpose of the work was to study the effect of the liquid water-soluble feed additive of complex action "Reasil®HumicHealth" (produced by LLC "Life Force" on the basis of a concentrated solution of high-molecular sodium salts of humic acids from Leonardite) on the vital indicators of the state of the body of broilers during rearing, their slaughter and meat qualities, dynamics of morpho-biochemical and immunological parameters of blood and micromorphometric characteristics of the organs of the immune system of poultry, as well as on organoleptic indicators and the yield of culinary products from poultry meat.

2. Medical and biological, chemical and functional properties of preparations of humic acids

Humic acids (from Latin humus - earth, soil) were isolated from peat. They have been the object of scientific research by scientists from different countries for over 200 years. Currently, natural substances that are at different stages of humification, located in different parts of the biosphere, have become the sources of obtaining humic preparations. These are compost, peat, brown coal, and sapropel. In this connection, the term "humic substances" is associated with substances of natural origin. It is believed that the formation of humic substances or humification is one of the largest processes of organic matter transformation. Some of the dead remains are mineralized to CO_2 and H_2O, the rest turns into humic substances [1].

Humic acid is a long chain of molecules that, in complex combination with fulvic acid, form a bioavailable complex for the healing of a living organism. Its value is due to the presence of more than 70 different components of minerals, amino acids, vitamins, natural polysaccharides, sterols, hormones, fatty acids, plant pigments, natural antioxidants, etc. Such a concentration of biologically active substances determines the variety of positive effects of humic acids on living organisms [2].

Tests of preparations of humic acids revealed that they do not exhibit carcinogenic, allergenic and anaphylactogenic properties. Humic preparations are not teratogenic, that is, they do not disrupt embryonic development, and do not cause morphological anomalies and malformations in animals. These preparations are not embryotoxic - their use does not provoke intrauterine death, a decrease in the weight and size of the embryo. This allows one to classify them as harmless to animals and humans, which provides significant advantages over classical preparations. Due to this fact, it is possible to produce environmentally friendly natural feed additives on their basis, as well as veterinary preparations for poultry, farm animals, fish and pets.

Using radioisotope labeling, it was proved that animal products do not contain residues of humic acids, which means that they are not absorbed into the blood and lymph, but have therapeutic effect in the gastrointestinal tract and intestinal walls. The healing and prophylactic properties of humic acids are based on their ability to envelop the intestinal mucosa of animals and to reduce or completely prevent the absorption of toxic metabolic products after infection, as well as when feeding poor quality feed. Humic acids are simply mixed into the feed, they are tolerated and do not have any side effects on the animal organism.

In the treatment of intestinal diseases, there is a decrease in pathological impulses from the peripheral nerve endings of the intestine, as well as the restoration of normal peristalsis and tone. Under the influence of humic acids, intestinal immunity is restored in animals exposed to stress, and under a slight tanning effect, the intestinal mucosa thickens, its permeability and excessive secretion of tissue fluid into the intestinal lumen decrease. Thus, dehydration of the body is prevented [2].

Scientists have accumulated extensive material on the effect of preparations with humic acids on animals' immune status. Humic acidsstimulate the body's immune system to defend against foreign influences. Under the influence of humic acids, the phagocytic function of leukocytes is enhanced, the body's defenses are additionally stimulated, and this reduces mortality and improves the safety of young animals [3].

Humic acids supply micronutrients, enriching the immune system, which enables animals to resist disease effectively. In addition, they inhibit the growth of pathogenic bacteria and molds, reduce mycotoxin levels, and improve protein digestion and the absorption of calcium, trace elements and nutrients. The result is high fatness, as well as immunity to disease [4].

The antiviral action of humic acids is considered to be the most effective in the therapy of animals, since the immunomodulatory effect of the preparation plays an important role in recovery. R. Laub (2000) found out that polyphenolic compositions based on humic substances have antimutagenic and antiviral effects.

The high biological activity of preparations with humic acids is also manifested in relation to fungal diseases. In particular, they have a fungicidal effect on *Candida albicans* inhabiting the gastrointestinal tract of animals and humans.

The varied composition of organic acids in humic acid preparations helps to break down food particles in addition to the action of enzymes, thereby having a positive effect on digestibility and feed conversion. This leads to an increase in productivity and in livestock production [5].

A large body of researchshows the absolute safety of humic acids for animals, humans and the environment. Their positive therapeutic effect on almost all types of animals has been proven. The introduction of humic acids into the diet of animals and poultry leads to the activation of their vitality, rapid adaptation to changing environmental conditions, acceleration of feed fermentation due to the development of beneficial microflora of the gastrointestinal tract, and to growth-stimulating and immunomodulatory effects [6].

Back in 1991, T.D. Lotosh noted that preparations based on humic acids can be used in medicine and veterinary medicine as a nonspecific preparations that increases the body's resistance to various adverse factors. The first medical preparation based on humic acids registered in Russia was "Gumisol" - a 0.01% solution of humic acids in isotonic sodium chloride solution [7].

The strengthening of the immune system is evidenced by the fact that humates, forcing the immune system to recognize its own dead cells, thereby reduce the level of infection [4]. The reconstitution of the body's immunity is facilitated by the additional transfer of iodine from foods to the thyroid glands under the influence of huminates [8].

Humic acids in medicine, veterinary medicine, and animal husbandry are considered to be means of increasing the body's resistance to the action of various unfavorable factors. The adaptogenic effect of humates is manifested at the cellular level. In particular, there is a stabilization of the values of the mitotic index, a decrease in abnormal figures of mitosis, an increase in the DNA content to the control level [9]. In Poland, a natural immunomodulator is produced. It consists of many components, including humic acids, which has an interferonogenic effect and is an inducer of tumor necrosis factor [10].

Scientific research allows one to conclude about the antiviral activity of preparations based on humic substances. Humic acids prevent viral replication by acting on the viral protein and block the absorption of viral particles on the cell surface. It is believed that molecules of humic acids encasethe virus in a kind of shell, block its entry into the cell and prevent replication. The antiviral effect of humic acids is more pronounced than the antibacterial, since have immunomodulatory effect on the host organism. The antiviral properties of humic acids are manifested against influenza A and B viruses, Coxsackie A_9 virus, herpes, rhinoviruses, ARVI and even against human immunodeficiency virus [11–15].

3. Development of "green technology" for rearing and fattening of broiler chickens in a production environment

According to experts, the leading place in the development of the world economy in the 21stcentury will be occupied by ecological technologies, colloquially called "green technologies". According to the classification of the Organization for Economic Cooperation and Development (OECD), one of the main purposes of "green technologies" is to eliminate the use of harmful synthetic chemicals in agriculture, as well as to introduce new environmentally friendly technologies in agriculture, animal husbandry and agricultural processing.

One of the stages of environmentally friendly technologies for the production of livestock products is the improvement of livestock and an increase in the productivity of animals and poultry with the help of feed additives, which are natural complexes of biologically active substances with the properties of antioxidants, adsorbents and antibiotics. This trend has led to an increase in the research on the use of alkaline salts of natural humic acids in animal husbandry [16].

The so-called quinones, which have antibiotic properties, have been found in the biological complex of humic substances. Therefore, preparations of humicacids are considered to be a promising alternative to antibiotics and probiotics in stabilizing the intestinal microflora, which no less successfully neutralize pathogenic microflora, while suppressing inflammation and blocking the sites of adhesion of pathogenic pathogens in the intestinal mucosa. Humic acids bind pathogenic *E. coli* by an average of 94%, and endotoxins by 82%. Bacteria and toxins bound by humic acid are clearance naturally [3].

In comparison with antibiotics, the mechanism of humic acid therapy manifests rather slowly, within 24–72 hours. In this case, pathogenic microflora is excreted, stimulating the formation of antibodies, and increasing the resistance and defense systems of the animal.

Veterinary control over preparation residues in livestock products is of great importance in production of environmentally friendly antibiotic-free products. Research by K.V. Korsakov and S.V. Kozlova (2019) evidenced that the combined use of the antibiotic "Florfenicol" and the feed additive "Reasil®HumicHealth" based on humic acids in a dose of 1 g/kg of feed accelerates the elimination of the antibacterial preparationby 2 times after the cessation of antimicrobial therapy. Use of the "Reasil®HumicHealth" in a dose of1.5 and 2 g/kg of feed acceleratesits elimination by 4 times after the cessation of antimicrobial therapy. Thus, the preparation of humic acids "Reasil®HumicHealth" is able to bind the metabolites of florfenicol and accelerate the elimination of the antibacterial preparation by 2–4 times, depending on the dose of the additive [17].

The essence of the interaction of a living cell with humic acids is that intact molecules of humic acids and high-molecular residues of their intracellular digestion are localized in the cell walls or in the layer adjacent to the cytoplasmic membrane. A semblance of an active filter appears on the surface of a living cell, which binds heavy metal ions into stable chelate complexes, intercepts molecules of pesticides and other organic xenobiotics, and also binds free radicals formed in the plasma membrane as a result of lipid peroxidation.

With this interaction, the release of energy is noted, which, instead of being spent on compensating for the adverse effects of the external environment, is used by the cell itself for growth and reproduction, which ultimately leads to an increase in its competitiveness [1, 18].

Due to their chemical properties, humic acid preparations bind heavy metal cations, compete with conventional mineral adsorbents (activated carbon, clay) and have adsorptive properties. In this case, humic acids slip between the villi of the intestinal epithelium and create a protective film from the finest particles of humic acid, which protects the tissues of the epithelium and lymph glands.

The adsorptive effect of humic acids is enhanced by their ability to penetrate into the small intestine without changes and to show their ability in the place, where toxic substances are fixed, their absorption slows down and excretion with feces is accelerated.

The high abilities of the natural-organic complex of humic acids "Reasil® HumicHealth" to sorb and desorb 5 mycotoxins of mixed feed - aflatoxin B_1, ochratoxin, toxin T-2, zearalenone and fuminisin B_1 have been experimentally established. At the same time, the feed additive is effective both at the maximum permissible level of mycotoxins in the compound feed, and at the concentration of toxins exceeding 5 MPC [19].

The hepatoprotective ability of humic acids was observed in experiments on broilers using rapeseed meal containing an increased amount of secondary plant metabolites that could cause liver damage. The low AST level in the blood of chickens fed with humic acid indicates its hepatoprotective ability [20].

Use of liquid Reasil®HumicVet in feeding broiler chickens of the Cobb 500 cross at the age of 9 to 34 days and laying hens of the High-Line cross during the period of intensive egg laying in an amount of 0.06 ml per 1 kg of live weight contributed to an increase in productivity and the safety of poultry, an improvement in the morphological state of the liver and a decrease in AST and ALT indicators. This indicates the restoration of liver functions and helps to prevent fatty and toxic degeneration and hepatosis [21].

4. Slaughter and meat qualities of broiler chicken carcasses, depending on the concentration and method of administration of humic acid salts

The productivity of modern crosses of broiler chickens largely depends on the technology of poultry operation, including the use of biologically active additives in feeding, which can have both a positive and a negative effect on the digestion and absorption of nutrients in the diet, which accordingly affects the quantity and quality of meat products.

To characterize the slaughter qualities of broiler chickens, a control slaughter of poultry with an average live weight in a group and anatomical and morphological carving were carried out. A set of the following indicators was assessed: the weight of carcasses, the yield of semieviscerated carcass (without blood and feathers, the intestines with cloaca were removed, the filled goiter, oviduct in females) and carcasses of complete evisceration (all internal organs were removed, the head between the second and third cervical vertebrae, neck without skin at the level of the shoulder joints, legs at the tarsal joint).

Analysis of the results of slaughtering poultry at the age of 31 days (**Table 1**) showed that broilers of the 1st and 2nd experimental groups, who ate a dry feed additive based on unmodified microporous humic acids "Reasil®Humic Health", at a dose of 1 and 1.5 g per 1 kg of compound feed exceeded their counterparts from the control group in terms of the weight of semieviscerated (by 41.34 and 36.34 g) and eviscerated carcasses (by 51.66 and 61.0 g). The yield of carcasses in the groups was, respectively, 84.62 ... 83.98% and 71.17 ... 71.60% versus 81.47 and 67.33% in the control.

Indices of the 2nd experimental group, according to the analyzed parameters, retain its position until the end of rearing (practically at the same level with the control).

Currently, the topical issue is the organization of production of high-quality natural poultry meat and bone semi-finished products. To study the development of meat forms of carcasses of broiler chickens, eviscerated carcasses after maturation in a refrigerating chamber (0 ± 4 °C, 24 hours) were dismembered into following parts: breast - lophosteon with ribs and muscles adjacent on both sides; leg - thigh and lower leg with attached muscles; shin - the tibia with adjacent muscles; thigh - femur with attached muscles; wing - humerus, radius, and ulna with adjacent muscles; skeleton - dorsal-scapular and lumbosacral parts of the carcass. Then the muscle tissue, skin, and bones were anatomically separated. The yield of a part of a carcass was determined by the ratio of the mass of the corresponding part of a carcass to the mass of a carcass of complete evisceration (%). The yield of muscle tissue, skin and bone tissue was determined by the ratio of the mass of a part of the carcass to the mass of the unknown tissue (%). The research results are shown in **Tables 2** and **3**.

It was found out that feeding the studied preparation based on humic acids in the composition of the compound feed of the experimental poultry does not have a negative effect on the weight and yield of the carcass components. Thus, the yield of the breast part of carcasses in broilers of the experimental groups, according to the results of their slaughter at the age of 31 days, increases from 31.81% in the control to 34.06–35.91% in the groups with humates due to a slight decrease in the yield of other components - leg and skeleton.

When broiler chickens were slaughtered at 36 days of age, their carcasses were characterized by the same yield of the compared anatomical parts in all experimental groups. However, with the subsequent slaughter of poultry, on the 38th day of the fattening, the ratio of parts of carcasses during carving changed.

In poultry carcasses that ate 1 g of humates per 1 kg of feed (1st experimental group), the percentage of the breast, leg and wing was almost equal to the control, however, the yield of the dorsal-scapular and lumbosacral parts (skeleton) decreased by 2.06%.

In the 2nd experimental group of broilers (1.5 g of the preparation per 1 kg of feed), there was an increase in the yield from the carcass of legs by 1.27, 0.79 and 1.53% in comparison with carcasses from the control, 1st and 3rd experimental groups, respectively. Such changes are primarily due to a decrease in the development of the breast of the carcass (yield - 37.27%).

The highest yield of the breast (40.11%) and forelimbs torso - wing (13.46%) during slaughter and carving was in the 3rd experimental, where broilers were fed with 2 g of the preparation with humic acids per 1 kg of feed.

However, the best results of the influence of humic acids on the intensity of development of the breast part of carcasses were during prolonged poultry feeding, that is, at the final stage of the fattening (42 days). During this period

Index	Control	1st experimental group	2nd experimental group	3rd experimental group
At the age of 31 days				
Preslaughterlive weight	1412.00 ± 89	1408.33 ± 77.98	1413.00 ± 79.05	1262.33 ± 102.22
Weightofsemieviscerated carcass, g	1150.33 ± 67.06	1191.67 ± 58.72	1186.67 ± 91.03	1024.67 ± 87.56
%	81.47	84.62	83.98	81.17
Weightofeviscerated carcass, g	950.67 ± 52.83	1002.33 ± 61.68	1011.67 ± 96.04	885.67 ± 67.54
%	67.33	71.17	71.60	70.16
At the age of 36 days				
Preslaughterlive weight	1782.50 ± 67.50	1734.00 ± 76.00	1729.00 ± 35.00	1681.50 ± 63.50
Weightofsemieviscerated carcass, g	1502.50 ± 36.50	1469.00 ± 90.00	1448.00 ± 34.00	1407.00 ± 92.00
%	84.29	84.72	83.75	83.70
Weightofeviscerated carcass, g	1305.50 ± 44.50	1271.00 ± 57.00	1251.50 ± 15.50	1210.50 ± 70.50
%	73.24	73.30	72.38	71.99
At the age of 38 days				
Preslaughterlive weight	1981.50 ± 61.50	1844.00 ± 79.00	1806.50 ± 63.50	1967.00 ± 145.00
Weightofsemieviscerated carcass, g	1712.00 ± 80.00	1557.00 ± 51.00	1549.50 ± 51.50	1680.50 ± 132.50
%	86.42	84.44	85.80	85.43
Weightofeviscerated carcass, g	1442.00 ± 67.00	1300.50 ± 54.50	1307.50 ± 25.50	1435.00 ± 151.50
%	72.79	70.50	72.38	72.95

Index	Control	1stexperimental group	2ndexperimental group	3rdexperimental group
At the age of 42 days				
Preslaughterlive weight	2402.00 ± 14.00	2197.00 ± 27.00	2139.0 ± 11.0	2511.0 ± 91.0
Weightofsemieviscerated carcass, g	2060.00 ± 14.0	1794.00 ± 39.0	1842.00 ± 6.00	2222.00 ± 10.0
%	85.76	81.66	86.12	88.49
Weightofeviscerated carcass, g	1774.00 ± 14.00	1530.00 ± 38.00	1577.00 ± 23.00	1910.50 ± 95.50
%	73.86	69.64	73.73	76.09

Table 1.
Slaughter qualities of experimental birds ($n = 3$).

Group	Weight of eviscerated carcass	Breast	Leg, including		Wing	Skeleton	Technical waste and process losses
			lower leg	thigh			
At the age of 31 days							
Control	920.0 ± 64.51	292.67 ± 35.07	125.33 ± 2.40	155.33 ± 2.91	100.67 ± 2.40	216.67 ± 13.68	29.33 ± 14.71
1st experimental	956.0 ± 40.84	343.33 ± 10.48	126.67 ± 9.68	141.33 ± 7.06	98.0 ± 7.57	201.33 ± 17.68	45.34 ± 4.16
2nd experimental	974.67 ± 86.35	332.0 ± 38.07	130.0 ± 10.58	150.67 ± 2.91	106.67 ± 6.57	226.0 ± 16.0	29.33 ± 14.71
3rd experimental	835.33 ± 66.03	284.67 ± 28.85	116.67 ± 8.74	118.0 ± 19.23	91.33 ± 5.70	194.67 ± 19.06	29.99 ± 10.26
At the age of 36 days							
Control	1305.5 ± 44.5	515.0 ± 5.0	171.0 ± 17.0	190.0 ± 6.0	131.0 ± 11.0	264.0 ± 4.0	34.50 ± 0.5
1st experimental	1271.0 ± 57.0	490.0 ± 56.0	164.0 ± 6.0	190.0 ± 0.0	135.0 ± 5.0	245.0 ± 15.0	47.0 ± 27.0
2nd experimental	1251.5 ± 15.5	476.0 ± 6.0	165.0 ± 9.0	187.0 ± 5.0	134.0 ± 6.0	256.0 ± 26.0	33.5 ± 15.5
3rd experimental	1210.5 ± 70.5	461.0 ± 27.0	167.0 ± 9.0	178.0 ± 8.0	125.0 ± 3.0	246.0 ± 28.0	33.0 ± 1.5
At the age of 38 days							
Control	1442.0 ± 67.0	569.0 ± 51.0	203.0 ± 3.0	259.0 ± 1.0	167.0 ± 15.0	233.0 ± 21.0	11.00 ± 7.0
1st experimental	1390.0 ± 42.0	542.0 ± 22.0	197.0 ± 1.0	255.0 ± 13.0	171.0 ± 5.0	196.0 ± 8.0	29.0 ± 12.0
2nd experimental	1414.0 ± 48.0	527.0 ± 1.0	207.0 ± 3.0	264.0 ± 4.0	173.5 ± 12.5	211.0 ± 4.0	31.5 ± 1.0
3rd experimental	1501.0 ± 115.0	602.0 ± 64.0	218.0 ± 12.0	259.0 ± 1.0	202.0 ± 36.0	206.0 ± 2.0	14.0 ± 2.0
At the age of 42 days							
Control	1798.0 ± 132.0	726.0 ± 108.0	230.0 ± 4.0	266.0 ± 22.0	188.0 ± 2.0	344.0 ± 2.0	44.0 ± 3.0
1st experimental	1545.0 ± 383.0	678.0 ± 182.0	177.0 ± 25.0	215.0 ± 51.0	159.0 ± 250	289.0 ± 83.0	27.0 ± 17.0
2nd experimental	1553.0 ± 23.0	705.0 ± 1.0	182.0 ± 10.0	217.0 ± 9.0	157.0 ± 5.0	283.0 ± 55.0	9.0 ± 7.0
3rd experimental	1919.0 ± 89.0	835.0 ± 45.0	252.0 ± 20.0	314.0 ± 6.0	237.0 ± 3.0	261.0 ± 23.0	20.0 ± 8.0

Note: $^*P \leq 0,95$to the control.

Table 2.
The ratio of parts of poultry carcasses of experimental groups, (n = 3).

Carcass part	Group			
	Control	1st experimental	2nd experimental	3rd experimental
Fattening period - 31 days				
Breast	31.81	35.91	34.06	34.08
Leg	30.51	28.03	28.80	28.09
Wing	10.94	10.25	10.94	10.93
Skeleton	23.55	21.06	23.19	23.30
Technical waste and process losses	3.19	4.75	3.01	3.60
Fattening period – 36 days				
Breast	39.46	38.55	38.05	38.08
Leg	27.65	27.85	28.13	28.50
Wing	10.03	10.62	10.71	10.33
Skeleton	20.22	19.28	20.46	20.32
Technical waste and process losses	2.64	3.70	2.65	2.77
Fattening period – 38 days				
Breast	39.46	38.99	37.27	40.11
Leg	32.04	32.52	33.31	31.78
Wing	11.58	12.30	12.27	13.46
Skeleton	16.16	14.10	14.92	13.72
Technical waste and process losses	0.76	2.09	2.23	0.93
Fattening period – 42 days				
Breast	40.38	43.88	45.40	43.51
Leg	27.59	25.37	25.69	29.49
Wing	10.46	10.29	10.11	12.35
Skeleton	19.13	18.71	18.22	13.60
Technical waste and process losses	2.46	1.75	0.58	1.05

Table 3.
Yield of carcass parts, %.

of the study the yield of the carcasses of specimen in the control was 40.38%, then in the carcasses of their analogs in the 1stexperimental group it was 43.88%, in the 2nd - 45.40%, in the 3rd group 43.51% or more, respectively, by 3.50; 5.02 and 3.13%.

The study of the morphological composition of carcasses of broiler chickens at the age of 31 and 42 days made it possible to establish the yield of various muscle groups: pectoral, femoral, as well as muscles of the lower leg and trunk, the ratio of edible (the mass of all muscles, skin with subcutaneous fat, liver, heart, muscular stomach, lungs, kidneys, internal fat) parts and inedible ones (head, legs, intestines with pancreas, bones, cuticle of the muscular stomach, glandular stomach, gallbladder, spleen, trachea, esophagus, goiter, thyroid and thymus glands, testes, ovary), the ratio of muscles to bones.

Index	Group			
	Control	1st experimental	2nd experimental	3rd experimental
At the age of 31 days				
Muscle yield to the eviscerated carcass mass, including	56.18	60.14	60.49	65.46
pectoral	28.05	31.25	30.66	26.31
femoral	10.11	11.64	12.04	16.61
muscles of the lower leg	7.06	7.76	8.30	8.92
muscles of the trunk	7.25	5.18	5.84	7.36
muscles of wings and neck	3.26	4.31	3.65	6.26
Ratioofediblepartstoinedible	1.09	1.40	1.36	1.32
Muscle-to-boneratio	2.08	1.99	2.05	1.92
At the age of 42 days				
Muscle yield to the eviscerated carcass mass, including	57.89	63.0	57.64	59.4
pectoral	29.88	40.30	30.79	31.47
femoral	11.88	7.94	9.54	11.08
muscles of the lower leg	8.50	6.94	7.12	7.07
muscles of the trunk	4.13	5.46	6.62	4.66
muscles of wings and neck	3.50	2.36	3.57	5.12
Ratioofediblepartstoinedible	1.47	1.93	1.67	1.88
Muscle-to-boneratio	1.75	2.24	1.97	2.08

Table 4.
Morphological composition of broiler chicken carcasses, %.

It was found out that the yield of muscle tissue in the carcasses of broilers of the experimental groups during slaughter on the 31st day of rearing had a significant difference, exceeding the value of the analogs of the control group by 3.96, 4.31 and 9.28%, respectively (**Table 4**).

The higher meat yield from the carcasses of broilers of the 1st and 2nd experimental groups is mainly due to the better growth of the pectoral muscles (31.25 ... 30.66%) and muscle tissue of the limbs (19.40 and 20.34%), in the 3rd group – of the femoral muscles (16.61%) and lower leg (8.92%), muscles of the wings and neck (6.26%). This was the result of the stimulating effect of the pretreated feed with the preparation with humates on the digestion processes and muscle tissue synthesis.

In the studied period of poultry slaughter the ratio of the mass of the carcass edible parts to the mass of inedible parts is 1.40; 1.36 and 1.32 respectively, which is higher than in the control group (1.09).

According to the results of poultry slaughter at the end of fattening, broilers in the experimental groups also exceeded the control ones, especially chickens that ate feed treated with a preparation with minimum humate content (1st experimental group). Their carcasses had high meat qualities - the ratio of edible parts to inedible ones was 1.93, muscles-to-bones ratio - 2.24, and the yield of pectoral muscles was 40.30%.

On the basis of experimental data, it can be stated that the inclusion of a dry feed additive based on unmodified microporous humic acids in the amount of 1 and 1.5 g per 1 kg of compound feed to the main diet of broiler chickens gives a more stable positive effect, both in terms of slaughter indicators (the yield of semi-eviscerated and eviscerated carcasses), and in the production of the most valuable natural semi-finished products (breast, legs) due to the intensive growth of muscle tissue.

5. Dynamics of morpho-biochemical and immunological parameters of blood and micromorphometric characteristics of the organs of the immune system of poultry using different concentrations of humic acid salts

The problem of natural resistance of poultry to diseases and unfavorable environmental factors has been studied by many researchers. Currently, interest in this issue is growing, which is explained by the need to know the features of the manifestation of the body's protective adaptations, that is, its resistance under different conditions of poultry rearing, feeding rations, etc. Nonspecific protective factors and adaptive capacity are in the body. It has a number of non-specific means of defense, neutralizing xenogeneic factors independently or in combination with each other [3, 22].

The purpose of the research was to study the dynamics of morpho-biochemical and immunological parameters of blood and micromorphometric characteristics of the organs of the immune system of poultry using different concentrations of humic acid salts. The functional and ontogenetic features of the first stage of poultry development were taken into account during the experiment.

According to the results of the studies, at 22 days of age, the morphological and biochemical parameters of blood in broiler chickens, both in the intact and experimental groups, were at the same level. So, the total protein level was 32 ± 0.4 g/l, the glucose level - 3.5 ± 0.05 mmol/l, the urea concentration was 3.7 ± 0.25 mmol/l, the hemoglobin level was 103 g/l, the number of leukocytes and erythrocytes - 26 × 109/l and 3.25 × 1012/l, respectively.

By the age of 35 days, a general increase in protein was observed in all study groups. The greatest increase was in birds of the 1st experimental group - 5 g/l. The glucose level increased in the intact group by 1.1 mmol/l, in the 1st experimental group by 1 mmol/l, in the 2nd - by 1.2 mmol/l, and in the 3rd group - by 1.1 mmol/l. The concentration of urea decreased and amounted to 1.2 mmol/l in the 1st and 3rd experimental groups, and 1.3 mmol/l in the control and 2nd experimental groups., In the 3rd experimental group the hemoglobin level decreased and amounted to 91 g/l, while in the intact group, this indicator was 87 g/l. The number of leukocytes was within the norm, and ranged from 21.2 × 109/l in the control group and up to 22.7 × 109/l in the 2nd experimental group. Compared with the 22nd day of the experiment, there was a decrease in the number of leukocytes. The number of erythrocytes decreased and amounted to 2.3 × 1012/l in chickens of the intact group, and in chickens of the 1st, 2nd and 3rd experimental groups it was 2.07, 2.07 and 2.32 × 1012/l, respectively.

By the age of 41 days, the highest total blood protein was in the 3rd experimental group (46 ± 0.4 g/l). The glucose level in all studied groups was the same and amounted to 5.5 ± 0.14 mmol/l. The concentration of urea, in comparison with 35th day, increased and amounted to: 1.4 ± 0.05 mmol/l in the control group, 1.3 ± 0.05 mmol/l in the 2nd experimental group, 1.1 ± 0.05 mmol/l and 1.03 ± 0.05 mmol/l in the 1st and 3rd experimental groups, respectively. The hemoglobin level increased by 10 g/l in chickens from the intact and 2nd experimental groups, by 18 g/l - in the 3rd experimental group.

The number of leukocytes in the blood of chickens of all studied groups increased. The greatest increase was in the experimental chickens of the 3rd group and amounted to 24.2 × 109/l. The smallest number of leukocytes was in the control group - 23.4 × 109/l. Compared to the 35th day of the experiment, the number of erythrocytes in control chickens increased by 0.05 × 1012/l and amounted to 2.35 × 1012/l, which was the smallest indicator of all studied groups. The largest number of erythrocytes, on 35th and 41st days of the experiment, was in the experimental chickens of the 3rd group - 3.15 × 1012/l.

Analysis of the research results showed that by the end of the experiment, the indicators of lysozyme and bactericidal activity of the blood serum of broiler chickens indicate a positive effect of different concentrations of humic acid salts on the indicators of nonspecific resistance of the poultry organism.

In chickens of the experimental groups, lysozyme activity of blood serum exceeded the analogs from the control by - 10.40, 12.09 and 12.28%, bactericidal activity of blood serum was higher than in the control group by 1.58, 1.76 and 1.79%, respectively.

When using the feed additive "Reasil®HumicHealth" based on humic acids at a dose of 1.5 and 2 g/kg of feed, indicators of lysozyme activity of blood serum of broiler chickens of the 2nd and 3rd experimental groups are statistically significant ($P < 0.05$) in relation to the control. Any statistically significant differences in the bactericidal activity of blood serum were not observed in all studied groups ($P < 0.05$).

An increase of hemoglobin by 12 g/l, erythrocytes - 0.8 × 1012/l, total protein - 4 g/l, leukocytes - 0.8 × 109/l, as well as lysozyme and bactericidal activity in all experimental groups indicates an increase in nonspecific resistance organism. The maximum dynamics was in the 3rd experimental group, where the additive was used at a dose of 2 g/kg of feed.

By the end of the experiment, the micromorphometric characteristics of the immune system organs of broiler chickens (by the age of 41 days) were characterized as follows. In the spleen of experimental poultry groups, in contrast to the intact, the average number and diameter of follicles exceeded by 11.94, 12.43, and 12.82%, respectively. The number and diameter of germ cents were higher by 30.18, 30.48, and 30.87%, respectively. In the factory bursa, in all experimental chickens, in contrast to the control group, the number and relative area of follicles were 28.6, 29.34, and 32.30% higher, respectively. The relative areas of the cortex and medulla of the follicles exceeded by 23.57, 27.16 and 30.32%, respectively. In the thymus of experimental chickens, in contrast to the intact ones, the relative area of the lobules of the cortical and medullary layers was greater by 18.46, 22.14 and 27.47%, respectively. The number of Hassal's bodies in experimental broiler chickens was 27.15% higher than in the control group.

6. Organoleptic characteristics and yield of culinary products from poultry meat

The taste qualities of the meat of the experimental poultry were evaluated by tasting, which allows one to reveal the influence of feed additives on it. The meat was subjected to mechanical (chopped semi-finished products) and heat treatment (cooking, frying. The smell (aroma), texture (hardness, tenderness), taste were determined. When developing culinary products, weight loss and the yield of the finished product were taken into account.

Analysis of the data shown in **Table 5** suggests that fattening broilers with feed based on humates increases the yield of culinary products from poultry meat after heat treatment.

For the periods of the study, the best results were in the production of chopped semi-finished products - chicken cutlets and croquettes, cooked by frying in the oven. This is probably due to the highest content in the poultry meat of the

Index	Group			
	Control	1st experimental	2nd experimental	3rd experimental
Fattening period - 31 days				
Boiled breast	55.41 ± 7.64	63.59 ± 6.84	52.50 ± 4.53	48.33 ± 4.64
Fried drumstick	54.63 ± 3.01	53.02 ± 7.81	50.79 ± 8.98	47.64 ± 11.55
Cutletfromminced mass	66.85 ± 3.25	74.81 ± 2.53	76.93 ± 1.04	78.29 ± 0.72
Ground-meatcroquette	94.86 ± 1.68	94.61 ± 0.47	95.65 ± 0.71	96.45 ± 0.45
Fattening period - 36 days				
Boiled breast	65.54 ± 1.24	66.01 ± 0.49	67.80 ± 11.5	64.95 ± 0.27
Fried drumstick	51.6 ± 2.26	54.59 ± 1.55	52.39 ± 0.16	52.23 ± 0.21
Cutletfromminced mass	63.75 ± 1.77	61.25 ± 1.77	60 ± 0.0	65 ± 0.0
Ground-meatcroquette	89.43 ± 1.49	93.08 ± 3.06	91.97 ± 1.49	88.01 ± 4.1
Fattening period - 38 days				
Boiled breast	69.28 ± 3.30	70.87 ± 1.46	69.63 ± 1.9	70.14 ± 2.88
Fried drumstick	56.08 ± 0.74	56.72 ± 0.14	66.27 ± 3.71	58.56 ± 0.09
Cutletfromminced mass	71.11 ± 1.57	79.47 ± 0.74	78.10 ± 2.69	86.95 ± 2.69
Ground-meatcroquette	92.13 ± 0.65	92.96 ± 0.52	94.94 ± 2.28	92.73 ± 0.19
Fattening period - 42 days				
Boiled breast	65.45 ± 1.23	71.11 ± 0.83	69.90 ± 1.8	66.61 ± 0.82
Fried drumstick	64.01 ± 0.40	69.98 ± 10.74	66.01 ± 2.02	63.21 ± 1.53
Cutletfromminced mass	64.86 ± 0.0	67.57 ± 3.83	67.57 ± 3.83	64.86 ± 0.0
Ground-meatcroquette	90.16 ± 0.23	90.31 ± 0.45	93.44 ± 0.16	90.48 ± 0.22

*Note: *P ≤ 0,90; **P ≤ 0,95 tothecontrol.*

Table 5.
Yield of finished products, %.

experimental groups of muscle tissue, which includes a protein capable of binding and retaining free moisture. However, it is necessary to take into account the fact that less weight loss when frying croquettes can be explained by the presence of fillers in the recipe, in particular bread, the starch of which partially binds the water released during the denaturation of meat proteins.

It should also be noted that the use of the studied feed additive in the amount of 1 and 1.5 g per kg of feed during 42 days of rearing and stockkeeping by the end of fattening contributes to the production of natural semi-finished products from carcasses - breast and drumstick. Their yieldaftercookingwas 71.11–69.90% and 69.98–66.01%, respectively.

During the tasting, there was a clear tendency: an increase in the concentration of humic acids in the broiler diet leads to a decrease in the taste of culinary products. It was found out that the maximum concentration of humates (2 g per 1 kg of feed) provoked the formation of an unpleasant metallic taste in products after heat treatment. However, the use of the additive at the rate of 1 and 1.5 g per 1 kg of feed does not have a negative effect on the sensory parameters of finished culinary products. They are similar in taste, color and texture to products of poultry fed with the basic diet.

7. Conclusion

Thus, based on the results of the studies, it has been established that the introduction of a liquid water-soluble feed additive of complex action "Reasil®HumicHealth" (produced on the basis of a concentrated solution of high-molecular sodium salts of humic acids from Leonardite) into poultry feed contributes to the development of sustainable safe processes of meat production of broiler chickens and finished products from it, forming the specified characteristics aimed at production of safe and high-qualified food products.

Author details

Inna Vladimirovna Simakova[1*], Alexey Alekseevich Vasiliev[1], Konstantin Vyacheslavovich Korsakov[1], Lyudmila Alexandrovna Sivokhina[1], Vladimir Vasilievich Salautin[1], Lyudmila Yurievna Gulyaeva[2] and Nikita Olegovich Dmitriev[1]

1 Department of Veterinary Medicine, Biotechnology and Food Technology, Federal State Budgetary Educational Institution of Higher Education Saratov State Agrarian University named after N.I. Vavilov, Russia

2 Department of Food Production Technology, Federal State Budgetary Educational Institution of Higher Education Ulyanovsk Agrarian University named after P.A. Stolypin, Russia

*Address all correspondence to: simakovaiv@yandex.ru

References

[1] Buzlama V.S. The mechanism of action of preparations of humic substances / V.S. Buzlama, V.N. Dolgopolov, V.N. Safonov, S.V. Buzlama // Results and prospects of the use of humic preparations in productive animal husbandry, horse breeding and poultry farming: collection of reports. Moscow, 2006. P. 24-33.

[2] Platonov V.V. Comparative characteristics of the structural features of peat humic and himatomelanic acids in relation to the specificity of their physiological action / V.V. Platonov, D.N. Eliseev, O.S. Polovetskaya, A.A. Khadartsev // Bulletin of new medical technologies. 2010.Vol. XVII. No. 4. P. 9-11.

[3] Koschaev, A.G. Features of poultry metabolism when using probiotic feed additive in the diet / A.G. Koschaev, S.A. Kalyuzhny, E.I. Migina, D.V. Gavrilenko, O. V. Koschaeva // Veterinary of the Kuban. 2013. No. 3. P.17-20.

[4] Ermagambet B.T. Effective use of humic preparations (based on humates) in animal husbandry and veterinary medicine / B.T. Ermagambet, E.V. Kukhar, N.U. Nurgaliev et al. // Achievements of science and education. 2016. No. 10 (11). P. 16-19.

[5] Eren M., Deniz G., Gezen S.S., Türkmen I. Effects of dietary humat on growth performance, serum mineral concentration and bone ash of broilers // Ankara ÜniversitesiVeterinerFakültesiDergisi. 2000. Vol.47. P. 255-63.

[6] Vasiliev A.A. Feed additives based on humic acids from Leonardite against mycotoxins / A. A. Vasiliev, K.V. Korsakov, S.P. Moskalenko, M.Yu. Kuznetsov, L.A. Sivokhina, I.A. Kitaev, V.E. Manieson // Feed production. 2018. No. 5. P. 33-37.

[7] Mashkovsky M.D. Medicines. Moscow, 1967. 708 p.

[8] Bard R. Effects of Humic Acid on animals and humans: An Overview of Literature and a Review of Current Research. 2002. P. 3-7.

[9] Gorovaya A.I., Oginova I.A. Molecular-cellular mechanisms of the adaptogenic effect of humic substances as a factor in the normalization of the state of cultural components of ecosystems // Ecological bases of reproduction of biological resources of the steppe Dnieper region: Collection of scientific papers. Dnepropetrovsk, 1986.

[10] Lotosh T.D. Experimental bases and prospects for the use of preparations of humic acids from peat in medicine and agricultural production / T.D. Lotosh // Biological Sciences. 1991. No. 10. P. 99-103.

[11] Inglot A.D., Zielinska-Jenczylik J. TolpaTorf Preparation (TTP) induces interferon tumor necrosis factor production in human peripheral blood leukocytes // Arch. Immunol. Ther. Exp. Warsz. 1993. Vol. 41, No. 1. P. 73-80; A method to assess the immunomodulating effect of the TTP by measuring the hypor activity to interferon induction and tumor necrosis factor response // Arch. Immunol. Ther. Exp. Warsz. 1993. Vol. 41, No. 1. P. 87-93.

[12] Klocking, R. Antiviral properties of humicacids.; et al. Experientia, May 1972, Vol. 28, Iss. 5, P. 607-608.

[13] Laub, R. The chemically induced inhibition of HSV infection. LaubBioChem Corp., August 1998. URL:www.laubiochem.com.

[14] Schols, D. Selective inhibitory activity of polyhydroxcarboxylates derived from phenolic compounds

against human immunodeficiency virus replication. Acquir Immune Defic Syndr, 1991; Vol. 4; Iss. 7, P. 677-685.

[15] Thiel, K.D., Heibig, B., Kloching, R., Vutzier, P., Sprossig, M., Schweizer, H. Comparison of the in vitro activities of ammonium humate and of enzymically oxidized chlorogenic and caffeic acids against type 1 and type 2 human herpes virus. Pharmazie, 1981; Vol. 36; Iss. 1; P. 50-53.

[16] Islam K.M.S., Schuhmacher A., Gropp J.M. Humic acid substances in animal agriculture // Pakistan J. Nutr. 2005. Vol. 4 (3). P. 126-134.

[17] Korsakov K.V. Influence of the feed additive Reasil®HumicHealth on the rate of excretion of antibacterial preparations from the body of broiler chickens / K.V. Korsakov, S.V. Kozlov // Poultry Science. 2019. No. 11. P. 78-84.

[18] Biryukov M.V. Biological action of humic acids on living cells // Peat in solving problems of energy, agriculture and ecology: Materials of the international conference. Minsk, 2006. P. 167-171.

[19] Vasiliev A.A. The value, theory and practice of using humic acids in animal husbandry / A.A. Vasiliev, A.P. Korobov, S.P. Moskalenko, L.A. Sivokhina, M. Yu. Kuznetsov // Agricultural scientific journal. 2018. No. 1. P. 3-6.

[20] Disetlhe A.R.P., Marume U., Mlambo V. Humicacid and enzymes inclusion in canola-based diets generate different responses in growth performance, protein utilization dynamics, and hemato-biochemical parameters in broiler chickens // Poultry Science. 2018. Vol. 97. Iss. 8. P. 2745-2753.

[21] Korsakov K.V. The use of hepatoprotective functions of the preparation of humic acids Reasil® HumicVet in poultry farming / K.V. Korsakov // Poultry Science. 2020. No. 9. P. 31-34.

[22] Bessarabov, B.F. Natural resistance and productivity of poultry / B.F. Bessarabov // Veterinary Medicine. 2010. 1-2 (86-87). P. 12-14.

Chapter 6

Humic Substances: Its Toxicology, Chemistry and Biology Associated with Soil, Plants and Environment

Rajneesh Kumar Gautam, Dimuth Navaratna, Shobha Muthukumaran, Amarendra Singh, Islamuddin and Nandkishor More

Abstract

In recent decades, scientists in different disciplines have been increasingly concerned about the fate of natural organic matter, and in particular of humic substances (HS). The term humic substances (HS) incorporates refractory autochthonous and terrestrial organic matter in the soil and aquatic ecosystem, and are one of the key fractions of natural organic matter. These substances are important chelators of trace elements constituting complex class of molecular structures that occur naturally, consisting of aggregation and assembly processes in which biomolecules derived from plant and animal residues are gradually transformed through biotic and abiotic tracts. Since these organic compounds are bound by or linked with soil mineral fractions, they must be physically or chemically separated from the inorganic components by an extraction method before their physico-chemical study. This chapter focuses on the chemo-toxicological, molecular aspects of humic compounds and their derivatives such as humins, fulvic acids, humic acids etc., with their agricultural, biomedical, environmental and biochemical applications. In addition to studying their impact on plant physiology and soil microstructure to expand our understanding about humic compounds.

Keywords: humic compounds, Humins, fulvic acids, humic acids

1. Introduction

Humic substances are a structurally specified fraction of soil organic matter that are thought to be an important component of the biosphere, as they are responsible for many physical and chemical processes in soil. Organic carbon accounts for approximately 62 percent of global soil carbon, with at least half of this carbon classified as the chemically resistant portion known as humic substances (HS) [1–3]. Plant and microbial biomass are the primary sources of organic matter production in soil, and their composition and properties are important major constraints for humification processes. When looking at the structural aspects of HS, we must understand the contributions of the different compound groups that make up those tissues. These compound types include intracellular (i.e. protein) and structural (for example, polysaccharides, lignin and aliphatic) compounds that shape membranes and cell walls, including membrane lipids and plant cuticles [4]. Due to their intimate relationships with soil mineral phases, they cannot be easily decomposed and are chemically complex to be used by microbes. One of the most notable properties of HS in soil is their ability to interact with metal ions, oxides, hydroxides, mineral and organic compounds [5], including toxic contaminants [3], to form water-soluble and water-insoluble complexes [4, 5].

Humic substances (HS) are important elements of natural organic matter (NOM) in soil, water, as well as geological organic substances including shales, brown coals, peat and lake sediments. They contribute to the brown or black color of surface soils and make up much of the characteristic brown color of rotting plant waste. They are major components of NOM in surface waters and can impart a dark color at higher concentrations, especially in brown fresh water ponds, lakes, and streams. The color of composts due to leaf-litter can range from yellowish-brown to black, depending on the degree of decay and concentration. It is well understood that HS are the most significant source of transient refractory organic carbon in the earth's crust. Conventionally, the mixtures are processed into humic acids (HA), fulvic acids (FA), and humin (Hu). HA is recognized as the fraction precipitated from an aqueous solution at $pH < 2$ (under acidic conditions), FA residues in solution after HA precipitation, and the Hu fraction is insoluble in aqueous solvents [6] .

2. Nature and origin of humic substances

Humic substances are soil complexes produced during humification phase, are responsible for fossil coal, oil deposits, and other geological deposits. It is the Earth's second most important organic phase, after photosynthesis. It is estimated that photosynthesis binds more than 5000 tons of atmospheric carbon per year, while the transformation of dead living organisms results in the creation of approximately 4000 tons of carbon [7], which is deposited annually on the Earth's surface. Humic substances are naturally occurring organic material that are formed due to bio-chemical, chemical, and physical changes in dead plants, microbial residues, and fauna debris. Humic material formation is linked to the behavior of microorganisms, resulting in the accumulation of recalcitrant macromolecular compounds with unique properties [7]. Due to death of microbes a recalcitrant humic mass is created due to their assimilation. These progressive decay leads to biotic and abiotic transformation processes which ultimately produces diverse and heterogeneous mixtures of polydisperse materials, which are a major component of natural soil organic matter. Humic material can take hundreds or thousands of years for their formation, resulting in a wide range of humic substances. Humic substances are extremely heterogeneous in composition, structure, and are derived from a variety of organic matter sources. This sets them apart from naturally occurring biomolecules such as proteins, carbohydrates, lipids, and others.

Humic substances happens to be the most common of the various organic substances found in nature. They account for approximately 25% of total organic carbon on Earth and can also be found in a variety of ecosystems such as lakes, rivers, sea sediments, soil, and coal composites. Humic substances influence the chemistry, cycling, and bioavailability of chemical elements in terrestrial and aquatic environments, as well as the distribution and degradation of xenobiotic and organic chemicals of natural origin [7]. For decades, scientists have discussed the formation mechanisms of HS [8, 9]. The lack of repeated sequences, as well as the variety of chemical and biological reactions involved in their formation, render HS very complex and multifaceted molecules capable of important signaling and nutritional functions in the soil–plant system. The polymeric origin of HS has a high molecular mass (100–300 kDa) [10], derived from lignin decomposition and abiotic catalysts such as primary minerals and layer silicates. Humic compounds are classified into three categories: (1) Humin, (2) Humic Acids (HAs), and (3) Fulvic Acids (FAs) [11]. **Figure 1** summarizes several of the major chemical characteristics of humic compounds.

Figure 2, illustrates the hypothetical model structure of fulvic acid (Buffle's model) [13], which contains aromatic and aliphatic structures, both extensively substituted with oxygen containing functional groups [15, 16].

2.1 Humus

Humus is generally brownish or black colored complex variable of carbon-containing compounds that does not show cellular organization in the form of animal bodies or plant under a light microscope. Humus is distinguished from non-humic substances such as carbohydrates (a primary component of carbon), fats, waxes, alkanes, peptides, amino acids, proteins, lipids, and organic acids by the fact that distinct chemical formulae for these non-humic substances can be published. The majority of small molecules of non-humic compounds are easily destroyed by soil microbes. Soil humus, on the other hand, decay slowly under natural soil conditions. When combined with soil minerals, soil humus can survive in the soil for centuries. Humus is the primary component of soil organic matter, accounting for 65 to 75 percent of the total. Humus plays an important role as a fertility factor in all soils, far outweighing its percentage contribution to total soil mass.

2.2 Humins

Humins are derived from humic compounds that are not soluble in alkali, not soluble in acids, and are insoluble at neutral pH ranges. However, the physical and

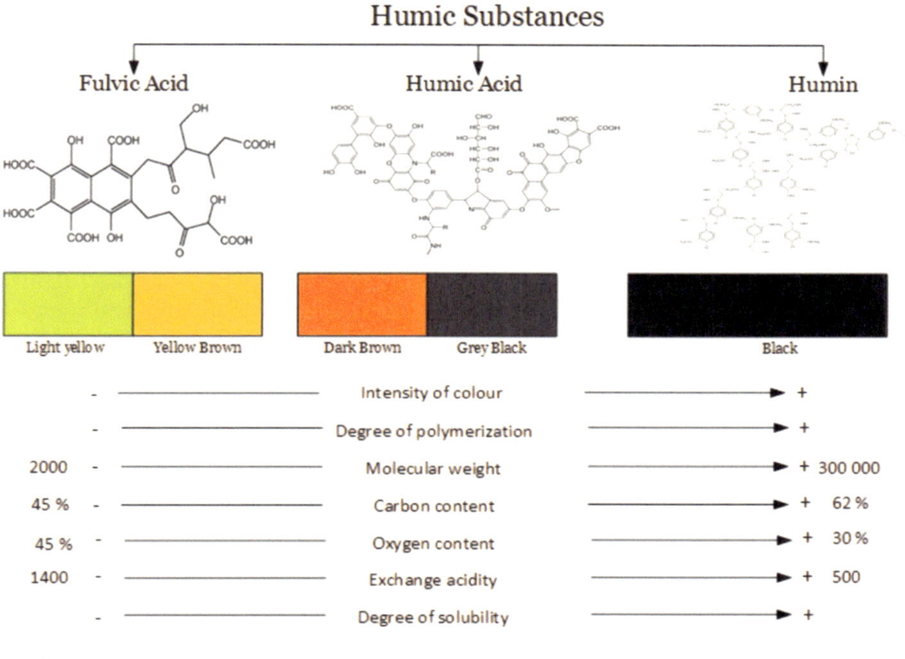

Figure 1.
Chemical characteristics of humic compounds (modified and adapted from [12–14]).

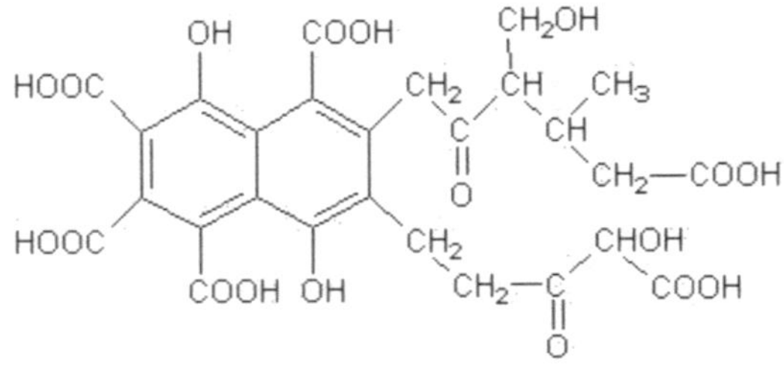

Figure 2.
Model structure of fulvic acid [13].

chemical properties of humins are partly known. Humin complexes are also recognized as macro organic substances due to their large molecular weights (MW) ranging from $1 x 10^5$ to $1 x 10^7$. Conversely, the molecular weights of carbohydrates range between $5 x 10^2$ and $1 x 10^5$ [17]. The humins are highly resistant to decomposition of all humic substances found in soil. Humins are known for enhancing the soil's water holding capacity, improving the soil fertility, taming soil structure, retaining soil stability, and acting as a cation exchanger.

2.3 Humic acids

Humic acids (HA) are organic acids that are a combination of weak aliphatic (carbon chains) and aromatic (carbon rings) organic acids that are insoluble in acidic water but soluble in alkaline water [18]. Humic acids are composed of humic compounds that precipitate from aqueous medium when the pH drops below 2. Due to their variable chemical composition, humic acids (HA) are also known as polydisperse. These complicated carbon-containing compounds are assumed to be 3-dimensionally flexible linear polymers that reside in the form of random coils with cross-linked bonds. On average, 35 percent of humic acid (HA) molecules are aromatic (carbon rings), with the remaining components being aliphatic (carbon chains) [17]. Humic acids (HAs) have molecular sizes ranging from $1 x 10^4$ to $1 x 10^5$. Clay minerals readily bind to humic acid (HA) polymers, forming stable organic clay complexes. Natural and synthetic-organic chemicals can be accommodated in the polymer's peripheral pores in a lattice (clathrate) arrangement. Humic acids (HAs) readily combine with trace inorganic mineral elements to form salts. It is reported in the literature that there exists more than 60 distinct mineral elements in naturally occurring humic acids (HAs). These trace elements are bound to humic acid molecules in a form that is easily assimilated by a variety of living organisms. As a result, humic acids (HAs) play an important role in ion exchange and metal complexing (chelating) [19].

2.4 Fulvic acids

The proportion of humic substances that are soluble in water at all pH levels. They are known as fulvic acids because they remain in solution after humic acid has been removed by acidification (FA). Fulvic acids range in color from light yellow to yellow-brown. The elemental composition of various FA and HA reveals that C, H, O, N, and S are the main elements in their composition. These major elements are always present, regardless of their birth, country of origin, or continent of origin. An examination of a broad range of HS reveals that the percentage composition of C, O, H, and N varies as follows: C (45–60),

Substances	% Dry ash-free basis			
	C	H	N	O
Lignin	62–69	5.0–6.5	—	26–33
Proteins	50–55	6.5–7.3	15.0–19.0	19–24
Humic acids	52–62	3.0–5.5	3.5–5.0	30–33
Fulvic acids	44–49	3.5–5.0	2.0–4.0	44–49

Table 1.
Composition of several plant material and humic substances (by Kononova) [16, 21].

O (25–45), H (4–7), N (10–13) and inorganic elements (ash) 0.5–5 **Table 1** [20]. summaries the composition of humic substances.

3. Preparation of fulvic and humic acids: a toxicological evaluation

The general properties for HA, and FA relate to their basic constituents (carbon, hydrogen, nitrogen and sulfur) and their oxygen-containing functional groups (hydroxyl, carboxyl, carbone and phenolic) which give rise to their overall chemical properties, including their ability to react with ionic particles in solution. Because of their properties, HAs and FAs have been used in a variety of agricultural applications, including improving nutrient and water use, soil quality, and carbon sequestration [11]. Although a variety of mechanisms have been proposed and/or studied for the effects on humic substances on plant life, much can be learned from the imperfect understanding of these mechanisms. One such process is ion exchange, which allows the use of plant minerals by prevention of the development of insoluble mineral complexes in the soil and then enhances humic substances in exchange for hydrogen and carbonic acid, which can produce absorbable mineral ions for plant roots. In sequestration of radioactive metals in soil, the ion-exchange process also works [22]. As humic substances play a role in maintaining plant life and their ability to attach or sequester potentially harmful environmental toxicants, they can also easily be extrapolated such that they can improve human nutrition and provide some protection from unintended ingestion of dietary toxic elements.

Two active New Dietary Ingredient Notices (NDIN) for use in a dietary supplement containing additional minerals to improve the status of minerals and trace elements were subjected to a specific preparation of HA and FA derived from Hungarian peat [20]. The NDIN has recorded 9 unpublished and one published clinical evaluations in which mineral status in humans has been improved and/or absorption is inhibited and toxic element excreted [23]. However, it is not clear if these effects can generally be extrapolated to humic substances due to the possible variations between this and other humic preparations as well as the inclusion of exogenous minerals. The use of clinical or mechanical research on other humic preparations, important for humans, is restricted and there are no other studies of the above-mentioned results, although it was shown that the addition of a humic extract to the broiler feed was intended to improve the growth of chickens. Due to the heterogeneous existence of the humic substances, an extrapolation for the whole community has not been deemed sufficient toxicological evaluation of one particular unique material, while attempts have been made for the identification of an appropriate model. As a result, a number of toxicological studies have been reported on humic materials of diverse origin even though many of them have been performed to examine the mutagenic potential of byproducts produced during water disinfection because of the normal occurrence of low levels of humic material in surface waters. Chlorination of organic non-volatile substances found in HA and FA is known to lead to the development in bacterial mutation tests of compounds (such as mucochlorine acid and 2, three, three-trichloropropenal) under conditions of decreasing pH and sufficient chlorine and HA and chlorine.

4. Plant physiology and humic substances

Various authors have assumed that even though high molecular weight (HMW) and small molecular weight fractions (LMW) appear to be behaving differently, HS may be adsorbed by the root plant [24]. The mechanisms by which HS interacts with the root cells affecting the plant physiology and growth have not been significantly researched until today. Changes in size and growth were the first to be analyzed among the modifications caused by HS on treated plants. HS will promote plant growth with a view to increasing plant length and dry and fresh weight under specific conditions. These effects are apparently dependent on the dosage and source of the drug, the plant species and age and the conditions of the trial cultivation [25]. Many studies have recently confirmed the assumption that HS has direct effect on plant physiology, especially in regard to the development of root hair and side root [26, 27]. The impact of HS has been thoroughly evaluated on metabolic processes in plants. For example, several studies have shown that HS affects breathing, protein synthesis and the function of the enzyme in higher plants [28, 29]. With regard to the photosynthesis process, few reports are available which focus on the contents of chlorophyll and transport of electrons.

4.1 Humic substances and abiotic stress regulation in plants

The productivity of crop plants will benefit significantly from regulation that mitigates the impact of abiotic stress on the growth and production of crops during ontogenesis. There are insufficient information currently available on control mechanisms and techniques. Optimizing plant growth conditions and providing regulators for water, nutrients and plant growth will help in avoiding abiotic pressure [30]. Bio-stimulants such as HS are also used in production systems with the aim to regulate physiological events in plants in order to improve productivity as well as these traditional strategies.

In the last two decades, scientists have increasingly utilized HS from natural materials [31]. In addition, HS provides new methods to alter physiological activity in plants to increase plant growth and abiotic stress tolerance. A number of examples show the ability of HS to improve plant abiotic stress tolerance. Treatment with hydrated leaves, anti-oxidants, radical growth and shooting under dry conditions, increased hydration of the leaves (*Agrostis palustris*) and increased fescue (*Festuca arundinacea*) [32]. Recent studies have shown that the blocking of cell wall pores by fouling has been responsible for the effects induced by HS that could be related to temporary mild water stress (a beneficial stress, 'eustress') and the downstream effects of HS on hormonal signals and plant growth [33].

4.2 Humic substances and improved mineral nutrition in plants

HS in the soil promotes root and spring growth through the improvement of mineral nutrition below the soil surface. In terms of plant production and active development, the function of these substances can be assessed [34], by their complementary and possibly diverse effects. HS regulates plant growth and mineral assimilation. In general, these effects are classified as direct and indirect.

HS activities rely mainly on the structural features, functional groups and their inclination to associate with inorganic, organic and soil substratum molecules [35]. Furthermore, HS primarily affects the bioavailability of nutrients through its ability to form complexes containing metallic ions, thus enhancing micronutrient availability (zinc, manganese, copper and iron). On the other hand, HS targets plant cell membranes that can cause biochemical and molecular processes at post-transcriptional levels in the roots and shoot [32]. Targeted HS typically increase macronutrient and micronutrient consumption in plants.

Jackson [36], used ^{14}C HS-listed substances, and found a small portion of them entering the root apoplastic pathway, especially those of low molecular weight. Therefore, the effect of HS to promote molecular systems and signal pathways of nutrient absorption present in cell membranes can be controlled. The full advantage of this direct influence is still unknown. Non-specific HS activity on the leaf and root can, however, be speculated on by controlling events at both the transcriptional and post-transcriptive stages, to modify molecular and biochemical processes.

5. Bimolecular fragments in soil humic substances

Humic soil fractions comprise bio-molecular fragments that were explicitly omitted from conventional humic concepts. Many of these bio-molecular movements are closely linked to the humic fraction by seven covalently and cannot be effectively isolated [37]. Why should a whole group of constituent molecules be excluded? Simply because it can be classified as biomolecules if we consider humic material as a super-molecular association of constituent molecules. It is inappropriate and deceptive for these nuanced, heterogeneous mixes to purge humic fractions of closely related, identifiable bimolecular components [38]. A description that is applicable to experimental studies of the humic component should include all molecules which are closely linked within a humic fraction, i.e. those which cannot be eliminated without substantial changes in the chemical characteristics of the fraction. That will necessarily include biomolecules that are identifiable and play important structural roles. Will this wider concept of humic molecules be contrary to the recognized refractory nature?

Soil humic compounds are typically aged between hundreds and thousands of years on average and many biomolecules easily decomposed by energy-seeking microbes and nutrients. However, it can lead to biomolecules being protected against microbial degradation by binding them with humic fractions, creating identifiable but refractory biological molecules [39], is alternatively a "two-part" view of the stability of humic substances. Although the majority of humic material is permanent, some humic elements may be transient. The persistent material contributes to ^{14}C, while the transitional components will contribute mainly to the C turnover rate. A significant portion of the transient, often refilled humic components may be made up of closely related, but easily degraded biomolecules. While most humic materials are permanent, some humic elements may be temporary. The continuous material contributes to ^{14}C, while the intermediate components are primarily responsible for the C turnover rate [40]. A large part of the frequently refilled transient humic components can consist of closely related, but easily degraded biomolecules.

6. Application of humic substances

Humus is one of the largest reserves of carbon in the world. To date, industrial uses of humus and goods extracted from humus have been rare. In the second half of the 19th century and the first half of the 20th century the use of coal was more abundant and, in addition, it was a foundation of chemical production. The application Petroleum was also considered the key raw material of 20th-century chemical industry. HS applications can now be broken down into four major categories: agricultural, industrial, environmental and biomedical.

6.1 Agriculture applications

From an agronomic and agricultural point of view, HS plays a significant role as they have a huge impact on soil quality and productivity. HS also shows a high Base Exchange potential, which is essential to soil fertility, in addition to enhancing the soil's physical properties and humidity. Many researchers and humates have observed the growth-promoting effect of humic substances in various preparations for plant growth improvement. Soil productivity in the presence of humic materials can be improved by various methods. However, in some case the consistency and fertility of the soil is greatly affected. In addition to the improvement of physical properties and moisture of the soil, HS shows a high Base Exchange potential, which is important for soil fertility. In various preparations for plant growth enhancement, several investigators have studied the growth-promoting effects of humic substances.

6.2 Environmental applications

Natural organic colloids are important because they form water solutions with a large number of metals, including radionuclides. It is well known that the humic substances affects the absorption of natural solids into the surface and ground waters of radionuclides. Thus, these organics may be essential as a transport agents in the environment for radionuclides. In environmental chemistry, the key role of humic substances is to extract radioactive metals, organic chemical substances of anthropogenic nature and other contaminants from water [18]. Calcium humate ion exchange products were considered suitable for the extract from the water and removal of radioactive elements from water discharges from nuclear power plants for heavy metals such as iron, nickel, mercury and cadmium and copper [11]. They also use their selective bonds to destroy ammunition and chemical warfare agents [41]. For the last 30 years, interactions between humic and micro-organic materials has been widely studied and researched. Fermenting bacteria have been found capable of reducing humic materials. Many investigators have been studying the use of humics as energy source for various bacteria, fungi and higher microorganisms, but they have not exploited them. Many researchers including [42, 43], investigated the use of humic as an energy source for specific bacteria, mushrooms and higher micro-organisms but pointed out that humic materials cannot be exploited as a source of food.

6.3 Biomedical applications

Veterinary and human medicine uses humic substances manufactured on a commercial scale. Various studies have been published on the medicinal properties of humic materials. The proliferation of gastric damage caused by ethanol was greatly reduced by humic acids administered prophylactically to rats [44]. TPP given to rats with gastric and duodenal ulcers improved the healing process considerably [44]. The interest in the use of humic materials in medicine and biology has developed in the past few decades. There is an investigation into the possibility that soil humus extract could be an applicant for cosmetic and pharmaceutical products with amino acid complexes and vitamin B analogues [11]. Anti-virals, profibrinolytic, anti-inflammatory and estrogenic activities are the primary explanation for the increased attention given to humic acids [45].

6.4 Chelate balneotherapy capacity of humic substances

In many mammalian cells, humic acid was, on the other hand, shown to be a toxic factor, although its precise cytotoxicity mechanism remains uncertain. Humic acid can be reduced from Fe (III) to Fe (II) under the wide range of pH values in aqueous conditions (4.0 to 9.0) and iron from ferritin reduction and release, but the process is partially inhibited by superoxide scavengers. It was shown that iron released from ferritin accelerates the lipid peroxidation caused by humic acid. Humic acids are also capable of reducing, releasing and supporting lipid peroxidation from ferritin storage. HAs coupled to freed iron can therefore disrupt the redox balance of the biological system and evoke oxidative stress. This may be one of the main pathways for cytotoxicity caused by HA [46]. Now is the time for new applications for humic substances, particularly in biomedicine, in less conventional areas.

7. Conclusions

HS has been widely studied in different fields of agriculture, such as ground chemistry, fertility and plant physiology as the major component of organic soil content. HS plays an important role in the regulation of pollutants' activity and mobility in the atmosphere and makes a major contribution to improving the status of global soil fertility. Together with a strong demand for safe food and sustainable agriculture, these features have led to increasing the environmental importance of

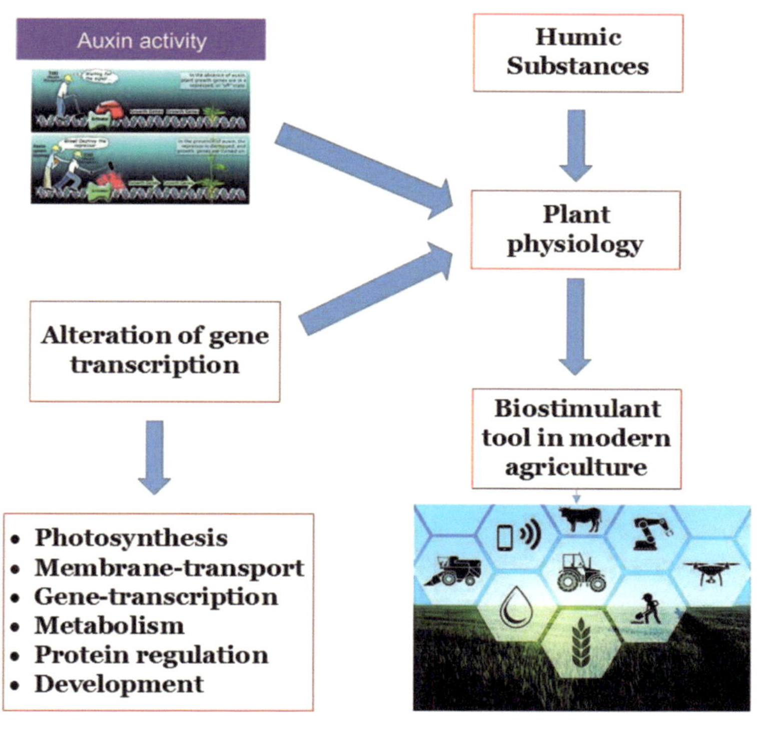

Figure 3.
Schematic representation of impact of humic substances on plant biology.

HS, recently recognized as a potential tool in facing global environmental problems. Many of its positive effects on soil structure and plant growth have been shown to depend on their chemical composition, but progress in HS research is known to be regulated by auxin. In addition, physiological and molecular evidence indicate that brassinosteroids may have a possible additional impact on the growth of plants. A critical review of literature shows that HS requires an auxin transduction pathway in order to determine its action in plant physiology but also show that the various signaling cascades in the global physiological response of plants to these substances (**Figure 3**). This could be used as a starting point for the clarification of mechanisms in the molecular response to HS in the plant. Further studies are required to evaluate the molecular goals and signaling trajectories involved in the intersection of the HS and the plant cells.

Author details

Rajneesh Kumar Gautam[1*], Dimuth Navaratna[1], Shobha Muthukumaran[1], Amarendra Singh[2], Islamuddin[3] and Nandkishor More[4]

1 College of Engineering and Science, Victoria University, Melbourne, Australia

2 Department of Civil Engineering, Institute of Engineering and Technology, Lucknow, India

3 Department of Civil Engineering, UIET, Babasaheb Bhimrao Ambedkar University, Lucknow, India

4 Department of Environmental Science, Babasaheb Bhimrao Ambedkar University, Lucknow, India

*Address all correspondence to: rajneesh.gautam@live.vu.edu.au

References

[1] Lal, R., *Soil carbon sequestration impacts on global climate change and food security.* science, 2004. **304**(5677): p. 1623-1627.

[2] Swift, R. and M. Hayes, *Humic Substances II. In Search of Structure.* John Wiley: Chichester, 1989. **449**.

[3] Orlov, D., *Humic acids of soils and general theory of humification.* Moscow State University, 1990. **1**: p. 325.

[4] Kögel-Knabner, I., *The macromolecular organic composition of plant and microbial residues as inputs to soil organic matter.* Soil Biology and Biochemistry, 2002. **34**(2): p. 139-162.

[5] Albers, C.N., et al., *Effect of different humic substances on the fate of diuron and its main metabolite 3, 4-dichloroaniline in soil.* Environmental science & techno logy, 2008. **42**(23): p. 8687-8691.

[6] Frimmel, F.H. and R.F. Christman, *Humic substances and their role in the environment*. 1988: John Wiley and Sons Ltd.

[7] Webe, P.J., *Humic Substances and their Role in the Environment*, in *EC Agriculture* 2020, EC AGRICULTURE Editor's Choice - 2020. p. 03-08.

[8] Burdon, J., *Are the traditional concepts of the structures of humic substances realistic?* Soil science, 2001. **166**(11): p. 752-769.

[9] Kelleher, B.P. and A.J. Simpson, *Humic substances in soils: are they really chemically distinct?* Environmental science & technology, 2006. **40**(15): p. 4605-4611.

[10] Schulten, H.-R. and M. Schnitzer, *Chemical model structures for soil organic matter and soils.* Soil science, 1997. **162**(2): p. 115-130.

[11] Peña-Méndez, E.M., J. Havel, and J. Patočka, *Humic substances-compounds of still unknown structure: applications in agriculture, industry, environment, and biomedicine.* J. Appl. Biomed, 2005. **3**(1): p. 13-24.

[12] Piccolo, A. and F. Stevenson, *Infrared spectra of Cu2+ Pb2+ and Ca2+ complexes of soil humic substances.* Geoderma, 1982. **27**(3): p. 195-208.

[13] Dalang, F., J. Buffle, and W. Haerdi, *Study of the influence of fulvic substances on the adsorption of copper (II) ions at the kaolinite surface.* Environmental science & technology, 1984. **18**(3): p. 135-141.

[14] Hayes, M.H., R. Mylotte, and R.S. Swift, *Humin: its composition and importance in soil organic matter.* Advances in agronomy, 2017. **143**: p. 47-138.

[15] Leenheer, J., et al., *Models of metal binding structures in fulvic acid from the Suwannee River, Georgia.* Environmental science & technology, 1998. **32**(16): p. 2410-2416.

[16] Weber, J., *Definition of soil organic matter.* Humintech: Humic acids based products, 2002.

[17] Pettit, R.E., *Organic matter, humus, humate, humic acid, fulvic acid and humin: their importance in soil fertility and plant health.* CTI Research, 2004. **10**: p. 1-7.

[18] Boggs Jr, S., D. Livermore, and M.G. Seitz, *Humic substances in natural waters and their complexation with trace metals and radionuclides: A review.* 1985.

[19] Pandey, A.K., S.D. Pandey, and V. Misra, *Stability constants of metal–humic acid complexes and its role in environmental detoxification.* Ecotoxicology and environmental safety, 2000. **47**(2): p. 195-200.

[20] Murbach, T.S., et al., *A toxicological evaluation of a fulvic and humic acids preparation.* Toxicology reports, 2020. 7: p. 1242-1254.

[21] Rice, J.A. and P. MacCarthy, *Statistical evaluation of the elemental composition of humic substances.* Organic Geochemistry, 1991. **17**(5): p. 635-648.

[22] Kapoor, A. and T. Viraraghavan, *Fungal biosorption—an alternative treatment option for heavy metal bearing wastewaters: a review.* Bioresource technology, 1995. **53**(3): p. 195-206.

[23] Klontz, K.C., et al., *The role of adverse event reporting in the FDA response to a multistate outbreak of liver disease associated with a dietary supplement.* Public Health Reports, 2015. **130**(5): p. 526-532.

[24] Vaughan, D. and B. Ord, *Uptake and incorporation of 14C-labelled soil organic matter by roots of Pisum sativum L.* Journal of Experimental Botany, 1981. **32**(4): p. 679-687.

[25] Trevisan, S., et al., *Humic substances biological activity at the plant-soil interface: from environmental aspects to molecular factors.* Plant signaling & behavior, 2010. **5**(6): p. 635-643.

[26] Canellas, L., et al., *Humic acids crossinteractions with root and organic acids.* Annals of Applied Biology, 2008. **153**(2): p. 157-166.

[27] Canellas, L.P., et al., *Humic acids isolated from earthworm compost enhance root elongation, lateral root emergence, and plasma membrane H+-ATPase activity in maize roots.* Plant physiology, 2002. **130**(4): p. 1951-1957.

[28] Nardi, S., et al., *Chemical and biochemical properties of humic substances isolated from forest soils and plant growth.* Soil Science Society of America Journal, 2000. **64**(2): p. 639-645.

[29] Vaughan, D., *The stimulation of invertase development in aseptic storage tissue slices by humic acids.* Soil Biology and Biochemistry, 1969. **1**(1): p. 15-28.

[30] Onaga, G. and K. Wydra, *Advances in plant tolerance to abiotic stresses.* Plant genomics, 2016. **2**: p. 1-12.

[31] Yakhin, O.I., et al., *Biostimulants in plant science: a global perspective.* Frontiers in plant science, 2017. **7**: p. 2049.

[32] Van Oosten, M.J., et al., *The role of biostimulants and bioeffectors as alleviators of abiotic stress in crop plants.* Chemical and Biological Technologies in Agriculture, 2017. **4**(1): p. 1-12.

[33] Olaetxea, M., et al., *Abscisic acid regulation of root hydraulic conductivity and aquaporin gene expression is crucial to the plant shoot growth enhancement caused by rhizosphere humic acids.* Plant Physiology, 2015. **169**(4): p. 2587-2596.

[34] Zandonadi, D.B., et al., *Plant proton pumps as markers of biostimulant action.* Scientia Agricola, 2016. **73**(1): p. 24-28.

[35] Garcia-Mina, J., M. Antolin, and M. Sanchez-Diaz, *Metal-humic complexes and plant micronutrient uptake: a study based on different plant species cultivated in diverse soil types.* Plant and Soil, 2004. **258**(1): p. 57-68.

[36] Jackson, M., *New root formation in plants and cuttings*. Vol. 20. 1986: Taylor & Francis.

[37] Frimmel, F.H., *Photochemical aspects related to humic substances.* Environment International, 1994. **20**(3): p. 373-385.

[38] Sutton, R. and G. Sposito, *Molecular structure in soil humic substances: the new view.* Environmental science & technology, 2005. **39**(23): p. 9009-9015.

[39] MacCarthy, P., *The principles of humic substances.* Soil Science, 2001. **166**(11): p. 738-751.

[40] Campbell, C., et al., *Applicability of the carbon-dating method of analysis to soil humus studies.* Soil Sci, 1967. **104**(3): p. 217-224.

[41] Talmage, S.S., et al., *The fate of chemical warfare agents in the environment.* Chemical Warfare Agents: Toxicology and Treatment, 2nd ed. Chichester, West Sussex, England: John Wiley & Sons Ltd, 2007: p. 89-125.

[42] Gaur, A. and K. Bhardwaj, *Influence of sodium humate on the crop plants inoculated with bacteria of agricultural importance.* Plant and Soil, 1971. **35**(1): p. 613-621.

[43] Sollins, P., P. Homann, and B.A. Caldwell, *Stabilization and destabilization of soil organic matter: mechanisms and controls.* Geoderma, 1996. **74**(1-2): p. 65-105.

[44] Abshenas, J., R. Kheirandish, and A.R. Salary, *Gastroprotective effect of mummy on induced gastric ulcer in rats.* Comparative Clinical Pathology, 2014. **23**(2): p. 305-309.

[45] Mishra, T., H. Dhaliwal, and P.K. Gill, *CHARACTERIZATION OF HUMIC ACID SYNTHESIZED FROM WHEAT BRAN THROUGH MICROBIAL STRAINS ASSOCIATED WITH SHILAJIT.*

[46] Ho, K.-J., et al., *Humic acid mediates iron release from ferritin and promotes lipid peroxidation in vitro: a possible mechanism for humic acid-induced cytotoxicity.* Archives of toxicology, 2003. 77(2): p. 100-109.

Chapter 7

Effect of Trace Elements Accumulation on Mangrove Ecosystem and Their Interaction with Humic Substances: The Case of Nickel and Iron

Juliano de Oliveira Barbirato, Natália C. Ferreira and Leonardo B. Dobbss

Abstract

Mangroves are areas of permanent preservation, but anthropogenic interference in this ecosystem (for example the launching of pollutants from industrial, mining, fertilizer by farmers, sewage) is increasing startlingly. Preserve and look for ways to bioremediate mangroves is fundamental, since these maintain the productivity of coastal ecosystems and is thus regarded as a natural nursery. The need to study the mangroves has been growing in recent years, particularly in respect to the environmental characteristics of this ecosystem. This chapter aimed to draw a parallel between the damage that can be caused by the trace elements nickel and iron on the mangrove ecosystem, more specifically affecting the nutrition of mangrove plants, in addition to showing possible effects of the interaction of these metals with humic substances of organic matter acting on the mitigation of stresses caused to the ecosystem under study. Through surveys of the information covered in this chapter, it can be observed that the presence of trace elements such as Iron and Nickel at high levels can cause eminent stress to the plant structure within the scope of its sedimentary physiology and biochemistry. It is necessary to subsidize further studies so that it is explicit and approved by the scientific community that, this environment, which is sensitive and important, the basis for various trophic levels, needs greater attention from government officials for its preservation, as well as the restoration of those many mangroves that are polluted by being close to urban places, receiving an exacerbated supply of pollutants.

Keywords: mineral stress, biostimulant, mangrove, plant mineral nutrition

1. Introduction

The Mangrove is a coastal ecosystem, in transition between terrestrial and marine environments, characteristic of tropical and subtropical regions [1]. Despite being considered an ecosystem associated with the Atlantic Forest, the Mangrove is made up of areas of little biological diversity, but of great functional diversity [2].

It is composed of highly specialized halophyte vegetation, adapted to fluctuations in salinity, subject to the tidal regime and characterized by colonizing muddy sediments with low oxygen levels, on poorly consolidated substrates [3].

It suffers from water input from both the continent and the marine environment, the Mangrove ecosystem is of recognized importance, having primary primary productivity, filtering, storing and making available organic matter and nutrients, with a fertilizing influence on the adjacent coastal waters [4]. The "tangle" of Mangrove roots captures, traps and stabilizes interstitial sediments, forming a natural physical barrier against the erosion mechanisms generated by waves, tides and currents, promoting stabilization of the coastline [5]. In addition, canopies of vegetation protect adjacent terrestrial environments from strong ocean storms and winds [6].

Underestimated in the past, the Mangrove was considered an inhospitable environment for a long time [7]. Until the mid-1970s, the progress of the coast was equivalent to clean beaches, sanitary landfills, ports and cultivation to take advantage of the land of the old mangroves, which would have future consequences [8]. Regarding the economic and social importance of this ecosystem, this approach was partly responsible for the construction of ports, spas, coastal highways, urban expansion and the release of raw sewage into its channels, considerably reducing its extent [9].

In more recent times, the mangroves were cut and grounded for the expansion of cities and the implementation of various enterprises [10]. Most of the vegetation of the Mangroves in the city of Vitória has been destroyed in the last decades due to several factors, among which the occurrence of landfills, real estate speculation, urban projects, slums, sewage disposal and deforestation stand out [11]. In different parts of the world and in an increasingly present way, in parallel with the degradation process, ways are developed to recover the impacted environments and even provide conditions necessary for the creation of new areas [12].

Due to this phase of expansion of recovery, there are now several techniques that may or may not include planting and the recovery of an ecosystem can be defined as the act of returning it, as far as possible, to a condition close to the original [13]. This management method is based on knowledge of the processes essential to the development and sustainability of the productivity of the system as a whole and not of its parts, therefore, it is necessary to acquire specialized knowledge about plants and animals, that is, it is necessary to know, first of all, the structure and functioning of the ecosystem [14].

Several Brazilian municipalities are already working with replanting, recovery and restoration of mangroves, however, there is still a need to investigate which factors interfere in the growth and development of these plants. One of these factors is probably the accumulation of heavy metals in the substrate, which combined with the massive presence of organic matter and its main component, humic substances, which is recognized as a potential plant biostimulant [15, 16], may be causing a greater accumulation of these elements, which in high concentrations become harmful to plants, leading to death [17]. Therefore, according to the above, studies on the interaction of the mangrove with other biotic and abiotic components of this ecosystem are necessary. Data of this nature will be essential when adapting restoration and recovery methods. In addition, municipal, state and federal public authorities will be able to more safely base possible interventions in this ecosystem, since the law states that it is their obligation to recover degraded areas.

The purpose of this chapter is to carry out a consistent approach to mangroves and the intrinsic effects of trace elements on the specie *Rhizophora mangle* L. in order to subsidize the understanding in the treated area for the scientific community.

2. Mangroves

2.1 Characterization, location and importance of mangroves

Mangroves are present on about 60–75% of the planet's tropical coast. These ecosystems play an important role in the carbon balance of coastal systems, with an export of terrestrial carbon [C] to the oceans (11%) about 15% of the total carbon that is deposited in current marine sediments [18]. Mangroves contribute about 10% of the terrestrial dissolved organic carbon when compared to other habitats and this large export of organic matter directly interferes with the food webs of coastal systems [19–22]. Recently, coastal ecosystems are being strongly impacted by natural climate change and human activities that affect their structure and food web, resulting in broad economic consequences. In addition, adjacent oceans provide considerable amounts of nutrients to coastal ecosystems and regulate nutrient dynamics [22, 23].

The mangrove can reach high levels of primary production. Mangrove litter provides trophic subsidy in adjacent coastal waters (the "outwelling hypothesis"), through a food web the debris is converted into a more palatable microbial biomass, which in turn acts as a dominant food source for the higher trophic levels [24]. In view of the economic importance of fishing in mangrove systems and adjacent waters, trophic dependence is a widely publicized function on mangroves and an important argument for their conservation [25].

The mangrove is an ecosystem that occurs in tropical coastal areas related to low, flat and estuarine regions, bordering lagoons, rivers or channels, waterlogged, brackish areas, being influenced by the tidal regime, but absent from direct wave actions [26]. All of these aspects influence the large deposition of sediments and organic matter. It is considered a link, as it is a transition area between marine, terrestrial and freshwater environments and is characterized by the high variation that occurs between these areas due to the regimes of both aquatic and terrestrial environments. The substrate has a pasty consistency, little compacted, swampy, rich in organic matter, little oxygen and subject to tidal regime [27]. These systems are generally young, as their dynamics produce constant changes in these lands, the result of constant advances and retreats of the tides [28–30].

Regarding their origin, the characteristic species of mangroves have records from the Eocene (period when angiosperm species began to occupy the land-sea transition areas) about 60 million years ago [31]. In order to have all this adaptation on the part of the plants in the system as currently, there was at that time an adaptive evolution of angiosperms from the end of the Cretaceous and beginning of the Eocene in such a way that the plant species started to adapt and tolerate high concentrations of salts in the sediment, whose first species were the genus Rhizophora and Avicennia resulting from continental drift. These species occur in practically all latitudes where there are mangroves [30, 32].

From the mangroves that have disappeared, it is possible to point out the formation of terraces of marine construction where beach sandstone and raised sandbank occur, dehydrated and consolidated by clay and humid cement. Mangroves are systems in constant dynamics, with some of them still in full expansion, with a constant movement of the horizon or superficial layer of the mangrove due to the withdrawal and sedimentary deposition [33–35].

The penetration of mangrove roots in deep regions allows the reduction of tidal currents leading to the accumulation of clay and sludge, components of the nutrient cycling process that sustains the high productivity in the mangrove system [36]. Sediments are characterized by being native or alien, accumulating fine fractions due to their low energy in the environment, justifying fluid retention [37].

Mangrove vegetation provides habitat for a variety of wildlife. A large number of mammals frequently visit these habitats, but few live permanently. Also a wide variety of birds and fish inhabit the mangroves, as well as shrimp, which use the mangroves as a nursery [38]. The mangrove oyster is found on the aerial roots of Rhizophora species (red mangrove). Dead leaves and branches of mangroves serve as a food source for microorganisms, which in turn form the food base for juvenile fish and shrimp [39].

The mangrove ecosystem constitutes a large portion, equivalent to 60–70% of the coastline in the tropical and subtropical regions of planet Earth [40]. However, it is found in 118 countries and its area reaches more than 137,760 km^2 [41]. The largest extension of mangroves is found in Asia (42%) followed by Africa (20%), North and Central America (15%), Oceania (12%) and South America (11%). About 75% of mangroves are concentrated in only 15 countries [41]. In the Americas, mangroves cover about 4.1 million hectares [40, 42].

Brazil is currently the third country with the largest extension of mangroves in the world with 968,963 ha, 7% of the world area, behind only Indonesia and Australia [41], occupying approximately 51% of the area in South America [2]. Brazil has several Environmental Physiographic Units along its coastline, which were classified by their similar characteristics of each environment taking into geomorphological aspects, sea currents, climate, etc. [43].

The mangroves in the State of Espírito Santo are distributed from the mouth of the Rio Doce in the extreme north to the Itabapoana River, bordering the State of Rio de Janeiro, occupying approximately 70 km^2 [44]. Some mangroves are poorly anthropized like that of the São Mateus River, others are considerably anthropized like that of Vitória Bay [45].

Among the six typical species of Brazilian mangroves, four are found in Espírito Santo: *Rhizophora mangle* (red mangrove), *Laguncularia racemosa*, (white mangrove), *Avicennia schaueriana* and *Avicennia germinans* (black mangrove). In addition, associated species are found, such as: *Conocarpus erectus* (button mangrove); *Acrostichum aureum* and *Acrostichum danaeifoliun* (mangrove fern) and *Hibiscus pernambucensis* (cotton-beach) [46].

2.2 *Rhizophora mangle* L

For this species to develop in unstable environments it has adaptations such as the so-called root-anchor, allowing its support in swampy environments influenced by the variation of tides. The specie *R. mangle* L. also known as red mangrove, belongs to the family Rhizhophoraceae which includes 16 genera and about 150 species [47]. In Rhizophora, the seed still germinates within the fruit attached to the mother plant and remains attached to it by the cotyledons that form a necklace within which the primordial bud of the seedling is included (**Figure 1A**). When the seedling (propagule = "pen") detaches from the mother plant, the cotyledons remain attached to it, thus, the structure that reaches the substrate consists of a large hypocotyl and the seedling [48].

One of the most striking features in *R. mangle* is the presence of structures that project in various directions around its stem towards the ground, called "aerial roots" or support, which are in fact special branches with positive geotropism, which form large number of roots when in contact with the soil, these special branches are rhizophores, that is, branches bearing roots, with negative geotropism (**Figure 1B**) [49]. *R. mangle* has opposite leaves, elliptical, glabrous, without glands; apopetal flowers, diclamids; glabrous, dark brown fruit; it blooms and bears fruit all year round and has the geographic distribution of America, West Africa and some islands of the Pacific [50].

Figure 1.
*Characteristics of red mangrove plants (*R. mangle*) - (A) Propagule with a well-defined abscission band; (B) Aspect of the rhizophore.*

2.3 Ecophysiology of mangrove plants

Mangroves are systems that generate interest for researchers in the field of plant ecophysiology, due to their peculiarities and adaptations, such as, for example, high rates of salinity, temperature, tidal variation, nutritional deficiency, unconsolidated sediment and physical–chemical factors [48, 51], adaptations with which were possible through different morphological, anatomical, physiological and biochemical mechanisms presented by these plant species and which increase their productivity [51].

In general, the mangrove develops optimally with low salinity, however, some species tolerate higher salinity than others. Excessive Na^+ and Cl^- ions can lead to imbalances in several plant organs, however these responses vary between species. Mangrove plants generally decrease the water potential of their tissues in relation to sediment, thus allowing water absorption, which is one of the first adaptive mechanisms to excess salts. With the increase in the concentration of salts in the sediment, the concentration of Na^+ and Cl^- also increases, but this same occurrence is not seen for K^+ [51].

The accumulation of ions leads to the destabilization of cell membranes and even induces oxidative stress through the formation of ROS (reactive oxygen species), there is partial loss of the granal cell, cytochrome B6f and dissociation of polypeptides directly related to photosystem II. These peculiarities affect the absorption of carbon, and the increase of biomass. Hypersalinity can still cause disintegration of chloroplasts and mitochondria, alter (reduce) the chlorophyll concentration, which reduces photosynthetic capacity, due to chronic photoinhibition, however in *L. racemosa* plants it was observed at 30 ppm salinity, an increase in the concentration of chlorophyll a and b. It also reduces stomatal conductance and mesophilic conduction, limiting photosynthesis by low CO_2 diffusion [51].

Plants that grow in a saline environment have two problems, osmotic regulation and ion toxicity [52]. Mangrove plants need to restrict water loss and keep their potential low in the leaf interior through adaptations such as leaf thickening, increased nitrogen retention and water use efficiency. Mangrove plants are considered to have the highest water use efficiency rates of all C3 plants, because they have low rates of stomatal conductance and low transpiration [51, 53, 54].

2.4 Adaptive mechanisms

Halophyte species are naturally more tolerant to salinity, increase photochemical efficiency and promote greater CO_2 assimilation when compared to other plants not tolerant to salinity. *L. racemosa* when compared to *R. mangle*, has a higher assimilation of carbon for being more tolerant to salinity [54, 55].

When it comes to adaptations due to the fact that they are subject to flooding, in the mangrove species the circulation of water eliminates the possibility of absorbing toxic substances such as hydrogen sulfide from the degradation of sedimentary organic matter. When flooded, the species deal with hypoxia (complete or partial) leading to a decrease in the rate of growth and root death in *L. racemosa*, for example. However, its biomass is mainly allocated in the aerial part due to the flooding and flooding conditions that affect the gas exchange system and nutrient absorption by the plants in the long term. However, floods may have little effect on mangrove plants, due to their capacity to store oxygen in their root aerenchyma while the tide is low [48, 51].

Increase was observed in the content of chlorophyll a and b and carotenoid pigments at the points of greatest environmental pollution in both *R. mangle* and *L. racemosa* plants [56]. Chlorophyll b and carotenoids act as photoprotective pigments, therefore, an increase in their content constitutes a response by plants to environmental stressors. Since the chlorophyll a fluorescence is almost exclusively originated from chlorophyll a molecules associated with photosystem II, it reflects the primary photosynthetic reactions in the thylakoid membranes [57]. Variations in chlorophyll a fluorescence emission are indications of changes in photosynthetic activity and reflect the effects of environmental stress on the photosynthetic apparatus [51, 56].

3. Mineral nutrition

The study of plant nutrition establishes the essential elements for the life cycle of plants, how they are absorbed, translocated and accumulated, their functions, requirements and disturbances when in deficient or excessive quantities. As for the nutrient, it is defined as a chemical element essential to plants, that is, without it the plant does not live. For a chemical element to be considered a nutrient, it is necessary to meet the essentiality criteria (direct, indirect or both) [58].

Mineral nutrition is considered a determining factor for rooting, considering its involvement in the determination of plant morphogenic responses, such as the formation of adventitious roots, as well as the modulation of length and density. In this way, it is necessary to consider the influence of the various nutrients in the adventitious rooting, therefore, it is necessary to analyze the role of each nutrient, particularly in each phase of the process [59]. Although these nutrients are equally important for plant production, there is a classification, based on the proportion in which they appear in the dry matter of the vegetables. Therefore, there are two major groups of plant nutrients (not considering carbon [C], hydrogen [H] and oxygen [O]): 1) macronutrients which are the nutrients absorbed or required by plants in greater quantities [nitrogen (N), phosphorus (P), potassium (K), calcium (Ca), magnesium (Mg) and sulfur (S) (expressed in g kg^{-1} of dry matter)]. This class can be further divided into primary macronutrients that are N, P and K and secondary macronutrients that are Ca, Mg and S. 2) micronutrients that are the nutrients that are absorbed or required by plants in smaller quantities [Iron (Fe), manganese (Mn), zinc (Zn), copper (Cu), boron (B), chlorine (Cl) and molybdenum (Mo) (expressed in mg kg^{-1} dry matter)] [60].

Most essential elements are absorbed in ionic form, from the soil solution. The way in which these ions are displaced depends on a number of factors such as the flow of water in the plant, density of the roots, in addition to the genetic characteristics of the species. In addition, this process undergoes abiotic interference, in which each species will respond in a different way. Minerals are absorbed from the capacity of ionic selectivity, and this selectivity is limited by osmotic factors and interactions with ions present in the soil. Another important point when talking about nutritional absorption is that it can occur under a concentration gradient. The absorption of an element also depends on the availability in the soil, for this reason, the assimilation of these elements is related to biogeochemical cycles [61].

In addition to the elements said to be essential to plant life, there are others considered beneficial and also the group of toxic elements. As for the beneficial element, it is defined as that which stimulates the growth of vegetables, but which are not essential or which are essential only for certain species or under certain conditions [61]. Silicon (Si) and cobalt (Co) are considered beneficial to the growth of certain plants, as well as Sodium (Na), nickel (Ni) and Selenium (Se). It should be noted that even a nutrient or beneficial element, when present in high concentrations in the soil solution, can be toxic to plants. However, it is considered a toxic element, which does not qualify as a nutrient or beneficial element. Thus, the toxic elements, even in low concentrations in the environment, can present a high potential for harm, accumulating in the trophic chain and slowing the growth and leading to the death of the plant. As an example of potentially toxic elements, we have aluminum (Al), cadmium (Cd), lead (Pb), mercury (Hg) and etc. [62].

4. Mineral stress

Mineral stress is defined as the sub-optimal availability of essential nutrients or the toxicity of nutrients or other elements (especially Al, Na, Cl, Mn, and other heavy metals), it is a primary restriction on plant growth in a large portion of the land surface [63].

Excessive intake of minerals can also have harmful effects on physiology, which has led researchers in recent years to acquire accurate data on the minimum requirements and toxic dosages of minerals present in food. The level of mineral content in vegetables depends on a number of factors including the genetic properties of the species, climatic conditions, the characteristics of the soil and the degree of maturity of the plant at the time of harvest. Thus, some nutrients essential to vegetables are involved in the response of plants to stress [64].

5. Responses of halophytic plants to stress of trace element

Halophyte plants are a great example of evolved species and adapted to hyperosmotic environments. These vegetables have evolved to adapt to adverse conditions, such as: high salinity, xerothermic environments, cold at seasonal temperatures and tolerance to the presence of toxic ions, mainly in the form of sodium and chloride [65].

Starting with the sediment, in the mangrove, the sediment acts as a deposit of nutrients arising, in part, from the release of domestic and industrial sewage and the use of agricultural fertilizers and pesticides [66]. In addition, considerable amounts of nutrients reach the mangrove ecosystems dissolved in seawater through the movement of the tides. Every characteristic present in that sediment can affect the physiological characteristics of a plant, such as the concentration of photosynthetic pigments, whose characteristics of the plant that can be altered by a variety of environmental factors, such as: salinity, irradiance, flooding, heavy metals, state nutritional, pollutants and therefore a potential indicator of adverse environmental conditions [51, 56, 67].

It has also been suggested that plants tolerant to salinity would be better adapted to deal with environmental stresses, including heavy metals. This performance is notably due to the plant's ability to accumulate metals in its tissues, triggering toxic metal detoxification mechanisms [68, 69].

Three distinct phases between plant responses to stress [70], in which a fourth was added after [71]:

- Response Phase (beginning of stress): alarm reaction or general alarm syndrome (SGA) (deviation from functional normality, decline in vitality and catabolic processes exceeding anabolism);

- Restitution phase (continuation of stress): resistance stage (adaptation, repair and reactivation processes);

- Final Phase (long-term stress): stage of exhaustion (very high stress intensity, overloading the capacity for adaptation, leading to chronic disease or cell death);

- Regeneration Phase: partial or complete regeneration of physiological functions when the stressor is removed and the damage is not very high.

The heavy metals released into the terrestrial environment tend to be concentrated in the soil and sediments, which become a large reservoir available for the roots of plants, which are very vulnerable to variations in concentrations of these elements. In aquatic environments these elements are available for both roots and shoots [72].

5.1 Iron stress conditions

Iron is an essential micronutrient for plants, involved in several fundamental processes such as photosynthesis, respiration, nitrogen fixation, DNA and hormone synthesis [73]. However, at high levels, this element can be toxic to plants [74].

This element is abundant in nature, reaching about 5% of the earth's crust. Despite this, iron toxicity is not normally a problem for plants, since most of this nutrient in soils is unavailable to plants [75]. In acidic and/or flooded soils, or in areas where Fe ore is being mined or being processed, increasing the concentration of Fe can lead to the excessive absorption of this metal, which quickly accumulates in the leaves and often reaching levels of toxicity [76].

Iron, when free and in excess within the cell, is capable of generating free radicals, such as the hydroxyl radical, through the Fenton reaction [77]. This radical is extremely toxic for cellular metabolism, being responsible for the oxidation of biological macromolecules such as proteins, nucleic acids and membrane lipids [78].

Studies [69], demonstrated that mangrove species that were evaluated, presented low iron bioconcentration factor in the leaf tissue. The redox environment and the high content of organic matter decrease the bioavailability of this element. In addition, the retention of this metal in the root suppresses its translocation to the leaves [69].

Plants, when they are under high levels of Fe and, or under reducing conditions, often form on the root surface the so-called "iron plate" in aquatic plants and in some terrestrial plants subject to flooding. The formation of these plaques is closely related to the oxidation of Fe^{2+} to Fe^{3+} and the consequent precipitation of iron oxides on the root surface [38, 79].

Due to the high adsorption capacity of the functional groups of Fe oxides and hydroxides, the iron plate can act as a barrier to the absorption of nutrients and toxic metals by adsorption and, or co-precipitation, interfering in the availability of these elements in the rhizosphere and its subsequent absorption and translocation to the aerial part [79].

Fe, although abundant in nature, normally does not reach toxic levels for most plants, since most of this nutrient in soils is unavailable, especially in neutral and, or alkaline soils [76]. In acidic and or flooded soils, or in mining or mineralization areas, there may be an increase in the concentration of available Fe, resulting in excessive absorption of this metal, often reaching levels of toxicity [76].

In the leaves, in high doses of iron, the appearance of yellow/brownish spots was observed, followed by the appearance of necrotic lesions, decreased leaf area and leaf abscission, probably resulting from the formation of oxidizing agents in the intercellular spaces that react with the components of the cell wall and plasma membranes [80]. The roots became darkened, indicative of Fe oxide deposition [81]. Less root formation was also observed and the roots became short, thin, rigid and brittle. The visible chlorosis observed in some leaves was proven by the decrease in the content of chlorophyll a. The reduction of chloroplast pigments has been associated with oxidative damage caused by the accumulation of reactive oxygen species (ROS) induced by toxic levels of Fe in leaf tissues [82].

In general, the excess of Fe induces P, K, Ca, Mg and Zn deficiency causing multiple nutritional disorders [83]. This interference of excess Fe may be associated with the precipitation of iron oxides on the root surface and the consequent formation of a ferric oxide crust (iron plate). The ability of plants to form these iron plates on the root surface is considered part of the mechanism used by plants to tolerate high levels of Fe^{2+} [84]. The formation of iron plaque, however, can lead to nutritional imbalance due to the high capacity of iron hydroxide functional groups to immobilize nutrients and metals by adsorption and / or co-precipitation inhibiting absorption, transport and/or use of other nutrients such as P, K and Zn [85].

The presence of organic matter can promote the availability of Fe, presumably through the supply of soluble complexing agents that interfere with fixation. When Fe is added to the soil in the form of a chelate, or when chelation occurs by soluble organic compounds, the concentration and gradient in the soil solution is generally higher than the concentration of Fe and non-chelated gradient and diffusion transfer is much larger [86].

The phenomenon of oxy-reduction is the most important chemical change on plants that are in flood areas, and oxidation–reduction or redox potential, which is a quantitative measure of the intensity of this change. The most drastic changes that occur when a soil is submerged and its Eh falls, reducing NO to nitrogen oxides, and Fe III to Fe II [87].

In an attempt to prevent iron deficiency, the plants developed two ways for its absorption: I - reduction of Fe^{3+} in Fe^{2+} through the acidification of the rhizosphere, caused by the proton extrusion by plasma membrane H^+-ATPases. This reduction of Fe^{3+} in Fe^{2+} is promoted by a specific protein, Ferro Chelate Reductase (FRO). After reduction, Fe^{2+} is transported by specific membrane transporters (IRT) into the cells [88]. II - Phytosiderophores (compounds that have a high affinity for iron) are secreted into the rhizosphere, where they join Fe^{3+} forming a chelate complex (Fe^{3+} -FS). This complex is transported by specific carriers known as Yellow Stripe (YS) into cells [89].

The ionic strength can influence the electrical potential and spatial structure of organic matter (OM) and, thus, affect its complexation with Fe. The complexation of Fe and Humic Acid (HA) is influenced by the ionic strength. From known values of ionic strength of freshwater, estuaries and marine environments, it is predicted that the complexation of Fe and HA occurs mainly in freshwater bodies, such as rivers and lakes. PH is one of the most important factors in the complexation of Fe and OM. The peak intensity of the Fe-HA complex was improved by increasing the pH from 4 to 6 and reached a maximum at pH 6. When the pH exceeded 6, the intensity of the complex of Fe-AH gradually decreased. The complexation of Fe and HA is strongly pH dependent due to the deprotonation of HA and hydrolysis of Fe. This can be assumed that Fe-HA complexes are formed mainly in freshwater bodies, where pH and ionic strength favor complexation. The Fe-HA complex in most freshwater bodies would eventually migrate to the estuary, and dissociate to release Fe at a higher pH. Within the estuary, the released Fe would be recomplexed with other strong binders from coastal water. HA, as an important dissolved iron transporter, can transport Fe from fresh water to estuaries [90].

5.2 Nickel stress conditions

Some researchers consider Ni to be an essential micronutrient in certain plant species, especially when grown in urea media, since Ni is a constituent of the urease enzyme [91], and its deficiency it leads to a reduction in urease activity in plant tissues such as soy, rice and tobacco, leading to an accumulation of excessive urea, making it phytotoxic [92]. Although the mechanisms of its phytotoxicity are still poorly studied, it is known that high levels of Ni in plant tissues inhibit photosynthesis and respiration. Symptoms of toxic effects are related to tissue damage, growth retardation, chlorosis and other symptoms specific to plant species [93].

Industrialization and urbanism are responsible for the increasing contamination of soils and water resources, with different toxic substances that represent a potential risk to human health and agriculture [94]. Heavy metals and metalloids constitute a special group of two pollutants due to their non-biodegradability, as well as rapid transport to the food chain. Ni is a significant environmental pollutant. In 2008, Ni was classified as "the allergen of the year" [95]. Generally, the majority of Ni released to the environment by human activities, especially those related to the raw material, is used in electroplating and metallurgy industries, as well as the application of mineral fertilizers that contain this metal [69, 96].

During the last decades, studies that discuss Ni toxicity have gained notoriety mainly due to their concentration in contaminated soil reaching 20 to 30 times (200–26,000 mg kg^{-1}) higher than the average (10–1000 mg kg^{-1}) found in natural soil [85]. Ni is unique among heavy metals, unlike Cd, Pb, Hg, Ag, and several other metals that have no known physiological function in plants. It is a constituent of a large number of enzymes, including ureases, which play an important role in seed germination and is necessary for the nitrogen metabolism of plants. Therefore, Ni maintains the appropriate cellular redox state and participates in several important metabolic processes [69].

Ni toxicity to plants has become a worldwide problem that also threatens sustainable agriculture. Ni can affect the transport of nutrients to the cells of the roots and consequently change the entire plant physiology, such as reducing the levels of chlorophyll, the photosynthetic activity of the leaf, leading to tissue damage and strongly reducing the production of plant biomass. At the biochemical level, Ni has a deleterious effect on the function of the membrane [70, 96, 97]. In chloroplasts, for example, it affects the light capture complex (CCL) and carotenoid values. It can also interfere with the electron transport chain and its intermediates (such as cytochromes b6f and b559) in leaves. By inhibiting key enzyme activities in the Calvin Cycle, Ni can also slow the dark reactions of photosynthesis. Additionally, there is growing evidence that Ni toxicity in plants is associated with oxidative stress as reflected by the increased concentration of hydroxyl, superoxide, nitric oxide and hydrogen peroxide radicals [67, 98, 99].

The excess of Ni induces leaf iron deficiency [100]. It was suggested that Ni could reduce the amount of Fe via competition with Fe^{2+} through the mechanism of absorption of the root cell membrane and the inhibition of its translocation from the root to the formation of the diaspore, causing over-accumulation of Fe in the roots. This situation may be responsible for the production of chlorotic leaves, especially between the veins [67, 101]. Another effect generated by the presence of Ni is its competition for binding sites, as they have ionic rays similar to Ca and Mg, and through this competition they can decrease the absorption of these macronutrients. This possible decline in the absorption of nutrients can also result in metabolic diseases and reduce the activity of the proton pump, thus reducing the energy required for the absorption of K [101].

Thus, the decline induced by Ni in photosynthetic activity and the deficiency of photosystem II can be explained by the adverse effect of excess of this metal on the electron transport chain, as a consequence of a decrease in chlorophyll synthesis. This is consistent with other studies that show that through the stress generated by excess Ni, the plant reduces photosynthetic activity, as this element is linked to the inhibition of primary photochemical processes [101].

Ni induces a reduction in total fresh weight without the concomitant effect on dry mass, suggesting a change in plant water status, which may be the result of decreased water absorption or increased water loss, both of which may occur due to damage to cellular structures [102]. In addition, Ni can promote a decrease in succulence and leaf density, which indicate real water deficiency [103]. However, in low concentrations, Ni is considered an essential element mainly due to its function as an irreplaceable component of urease [104].

An important point is the "browning" of the roots of mangrove seedlings treated with Ni, as also observed in roots of *Z. mays*. The darkening of root cells may indicate the oxidation of phenolic compounds to cytotoxic quinones, which is believed to be mediated by the reduction of an electron to semiquinone radicals that self-oxidize to form ROS [96, 105].

Works with *R. mangle* exposed to trace elements such as Ni are hardly found, due to little study in this area. The sensitivity of this plant to Ni is still poorly known, however, a study with *R. mangle* in an impacted environment and an non-impacted one comparing the results of both, it was concluded that the in spite of the morpho-anatomical changes observed, it is possible to state that this species survives without the considerable impairment of its leaf structure, in areas subjected to environmental impacts, differently from the results found in the present study, which allows us to consider our results to be positive [81].

Reduced growth and reduced biomass production are general responses of plants superior to the toxicity of heavy metals, which can be linked to loss in cell turgor, resulting in decreased mitotic activity and / or inhibition of cell elongation. Thus, the toxicity of Ni can cause growth inhibition of the aerial part and/or roots, as observed in several tested plant species, such as roots, and aerial part [106].

The symptoms of toxic effects are related to tissue damage, growth retardation, chlorosis and other symptoms [107]. The phytotoxicity of Ni is the result of its action on the photosystem, causing disturbances in the Calvin cycle and inhibition of electrical transport because of the excessive amounts of ATP and NADPH accumulated by the inefficiency of the biochemical phase of photosynthesis [108, 109].

6. Dynamics of organic matter in mangroves

In mangrove sediments, because of the intrusion carried out by the flood, there is a decrease in gas exchange between the soil and the air, since the rate of oxygen diffusion in water is slower than in the air, creating a reducing environment. Thus, the decomposition of organic matter will occur by organisms that do not use O_2 with a receptor, but NO^{3-}, Fe^{3+}, organic acids, SO_4^{2-} and SO_3^{2-}. Therefore, microorganisms more slowly decompose organic matter, which accumulates in the soil [110].

The organic matter accumulated in the soil is a source of nutrients in the mangrove, since the decomposition and mineralization of plant material is important in the nutrient cycling that occurs in the soil–plant-atmosphere system, allowing part of the carbon incorporated by biomass through the process of photosynthesis return to the atmosphere as CO_2 and the nutrients absorbed by the plants are again usable [111]. The mineralization of organic matter is influenced by factors such as soil redox potential, microbial activity, plant species, litter production, litter C/N ratio and tidal flooding [112].

The mangrove sediments are classified as Organosols and Gleisols, depending on drainage conditions. Organosols have an accumulation of plant remains with varying degrees of decomposition and restricted drainage conditions [113]. The Gleisols present an intense reduction of the iron compound in the presence of organic matter, with or without oxidation alternation and are permanently or periodically saturated with water [114]. The water saturation is due to the influence of the tides, which is considered as the main mechanism of entry of saline waters in the mangroves [115]. Tides are responsible for important characteristics of these soils, such as reduced redox potential, existence of high pH values and dynamics of elements such as sulfur and iron [116].

Because of the effects of flooding on mangrove sediments, there is a decrease in gas exchange between soil and air, as the rate of oxygen diffusion in water is slower than in air. This interference in the diffusion of oxygen generates a reducing environment, causing the microorganisms to have their metabolism slowed down, since O_2 acts as an electron receptor promoting the maximum efficiency of decomposition of organic matter. With this, the microorganisms start to use electron acceptors other than O_2 for the decomposition of organic matter, following the following thermodynamic sequence: NO^{3-}, Mn^{4+}, Fe^{3+}, $SO4^{2-}$, CO_2 (methanogenesis), N_2 and H^+ in a way slower and less efficient, resulting in the accumulation of organic matter in the soil [110].

In regions where mangroves predominate, it is common to see a combination of high levels of organic matter, arising from the greater supply of plant biomass, anaerobic conditions, source of reactive Fe and sources of $SO4^{2-}$, from sea water, both readily available, making mangrove soil an environment conducive to the occurrence of bacterial sulfate reduction (RBS) [117]. This RBS is the dominant form of breathing in mangrove soils, with iron and sulfur compounds being important elements in the biochemical cycles of these environments [118].

The dynamics of Fe and S in these environments are affected by the amplitude of the tides, which sometimes cause oxidative conditions and sometimes reduce the mangrove soils [119]. Vegetation also influences, as it oxidizes the rhizosphere region by translocating the oxygen absorbed on the soil surface to the root structures located on the subsurface [116]. Therefore, elements such as iron and sulfur, which are sensitive to changes in redox conditions, have their dynamics altered and may suffer an increase or decrease in the solid fractions of the soil and interstitial waters [116].

In the mangrove, the rate of denitrification and ammonification can be high, due to factors such as anaerobic condition combined with high levels of organic matter [116]. Under flood conditions, high levels of ammonium can accumulate, since this ion is relatively stable in anoxic conditions, predominating in most soils [120]. The levels of inorganic nitrogen in the mangrove soil are usually low because of the low addition of N to the soil, losses by tide, reduced transformation processes and efficient microbial assimilation [116].

The decomposition of organic matter in the mangrove is determined by the quality of the litter present in the soil [116], and the chemical composition of the leaves can accelerate or delay the decomposition [116]. The leaves of the genus Rhizophora have constituents that are more difficult to degrade than the leaves of Avicennia, due to their large amount of polyphenols [121].

The mineralization of organic matter in mangrove soils is carried out by aerobic and anaerobic microbial processes. In anaerobic mineralization, microorganisms use other electron acceptors to replace O_2, e.g. NO_3 - → Mn (IV) → Fe (III) → $SO4^{2-}$ → CO_2. According to this sequence, there is a gradual decrease in the redox potential and in the free energy of oxidizing agents for respiration. Thus, the use of acceptors other than O_2, causes the decomposition of organic compounds to be slower, accumulating in the soil [116]. Anaerobic mineralization in mangrove soils, associated with the high capacity for biomass production, makes this area an accumulator of carbon and, therefore, important in the global cycle of this element and in the context of climate change [122].

The process of decomposition of organic matter through aerobics occurs on the soil surface and in the small aerobic zones around the roots where there is oxygen and a few centimeters from the water column [123]. In addition to aerobic and anaerobic respiration, suboxide respiration can also occur in the degradation of organic matter in the mangrove soil [116]. Under sub-toxic conditions, iron oxides and hydroxides are reduced in the decomposition of organic matter, generating high levels of Fe^{2+}, which can precipitate in the formation of carbonates, sulfides, phosphate or new oxidation, promoting the synthesis of ferrihydrite, lepidocrocyte and goetite [116].

Because seawater has high concentrations of sulfate, the decomposition of organic matter in mangrove soils is predominantly attributed to the process of bacterial sulfate reduction (RBS) [124], resulting in the formation of sulfide. The sulfide generated in the RBS process can have several destinations, such as precipitation in the form of iron sulfides - example: Pyrite (FeS_2) was considered as the final product, and more stable in the RBS process and participate in redox reactions [125]. Consequently, most mangrove soils contain high levels of inorganic sulfur in the form, mainly, of FeS_2 and elemental sulfur (S_0) [126]. In iron-rich environments, the oxidation of iron in the decomposition of organic matter can be comparable, or even greater, than the reduction of sulfate as this can be impaired in the presence of other electron receptors such as O_2 and Fe^{3+} [116].

The presence of functional groups of organic matter in the soil can complex trace metals present in the soil solution, and thus decrease the toxic capacity of the metals [127]. However, to explain the effects of the possible decrease in toxicity caused by nickel in *R. mangle* seedlings, which was observed in the vast majority of parameters analyzed, one should not only take into account the role of SH on plant growth and plant development as a whole, but also taking into account the complexing capacity of SH, since organic matter is considered an important mitigator of the toxic effects of heavy metals because it has the ability to form insoluble complexes with these metals and other elements, making plants less available, enabling the cultivation of plants in areas with high levels of contamination [128, 129].

6.1 Organic matter; humic substances and their possible interactions with trace elements

Soil is the basis of the entire production process on the planet. It is an important and well-organized component of nature, adjusted to multiple functions of dynamic balance. Its composition is varied and dependent on factors and formation processes (biogeochemical), as well as on the handling and use to which it is submitted [130]. The organic matter content (OM) of the soil is dependent on many factors that exert its influence individually and together, such as: climate, soil texture, topography, drainage, vegetation cover and land use [131]. The process of formation of organic matter in the soil is initially a biological process and almost all the flora and fauna existing in the soil has a direct or indirect effect [132].

OM is a complex matrix formed during the microbial decomposition of plant and animal waste that exists in soil, groundwater and rivers and plays a vital role in the global carbon cycle [133]. This OM can be divided into two fractions. The hydrophobic (non-humic) group that contains aliphatic chains of C and N, including carboxylic acids, carbohydrates, tannic acids (TA) and proteins and a hydrophilic one, composed of humic substances (HS), aromatic carbon, phenolic structures, and conjugated double bonds.

HS can be classified in terms of its solubility in different pH ranges: humic acids (HA) comprise the fraction that is insoluble at a low pH; fulvic acids (FA) are low molecular weight compounds that are soluble in a larger pH range and humine is the fraction that is insoluble at any pH.

The greater the amount of C and N, the greater the degree of evolution of humic materials. Atomic C / N, H / C and O / C ratios are often used to monitor structural changes in humic fractions and to elucidate the structural formulas of humic substances from different sources [134, 135]. The C/N ratio indicates the degree of incorporation of N and the degree of humification in the structure of SH [136, 137]. The O/ C ratio is normally used to estimate the abundance of oxygen contained in functional groups [138] and the H/C ratio is used as an index of aromaticity and aliphaticity of organic matter [136].

The root architecture of *R. mangle* seedlings was altered by adding SH to the culture solution. The number of lateral roots (NRL) of *R. mangle* seedlings was changed by the presence of the best dose of SH (6.25% HS) in relation to the control treatment [139–141].

SH and metals have interactions that are influenced by several variables, such as chemical load and heterogeneity of the humic material. These interactions of metals with SH are closely linked to ionic strength and not only to concentration that are also bound to their binding functions [142].

Mangrove sediments are known to have a great capacity to accumulate metallic elements, which is attributed in part to their high content of organic matter, which can act as an agent for complexing trace elements. In addition, the mangrove sediments are subjected to sulfate reduction processes, which leads to the precipitation of sulfide minerals that are capable of retaining, for example, Co and Ni in marine sediments. In mangrove sediments, the complexation processes of trace elements, however, depend on biogeochemical gradients, salinity, redox potential and organic matter contents that are moved by the distance to the seashore, the magnitude of the tide, and the types of mangrove tree species. In addition, diagenetic reactions can operate on sediments as a function of time, due to tidal cycles, flooding seawater, as well as freshwater intakes. The redox cycle in mangrove sediments can thus have a significant impact on speciation of trace elements [143, 144].

On the other hand, the humate's ability to leach minerals and mobilize metals is well known, and is assumed to be an important pedogenic process. With this there is a reconstitution of organic complexes and stabilization of metals initially released silicates and oxides. Several metals, including Cu, Zn, Co, Ni and Fe, are mobilized in soils and lead to reducing conditions [145].

Oxygen atoms in HS are mainly present in the form of carboxylic and phenolic groups [146] that can play an important role in creating binding sites for certain inorganic species (such as polarizable metal cations) and tracking organic compounds (including pesticides and endocrine disrupters) present in the environment [147].

HS are defined as supramolecular associations of organic molecules of small molecular mass that hold together because of the hydrophobic bonding forces of van der Waals, dipole–dipole and hydrogen bonds [148]. HS can play an important role in the bioavailability of metals in soils and, thus, contribute to the mitigation of toxicity to plants. Assessing the influence of HS on the mobility of Cu and Zn during aerobic composting of sewage sludge, a study [149] was observed an increase in the distribution coefficient of humic acids (HA)-Cu and HA-Zn of 27.5 and 3.33%, respectively, suggesting that there was a reduction in the mobility and availability of these metals.

The interest in the chemical characterization of HS is based on its marked influence on the solubility and mobility of trace elements in the soil and on the complexation of nutrients [150]. The ability of HS to interact with metals is normally attributed to the high content of functional groups containing O, such as carboxyls (CO_2H), carbonyls (CO) and hydroxyls (OH) [151]. The main chemical elements affected by redox reactions are: C, N, O, S, Mn and Fe. In contaminated soils, arsenic (As), Se, Cr, Hg and Pb can also be added.

In anaerobic conditions, resulting from O_2 depletion, after flooding the soil, or even in anoxic microsites, such as inside water-saturated micropores, aerobic microorganisms drastically decrease their activity, become quiescent or die [152]. Optional and anaerobic microorganisms then proliferate, using carbon compounds as a substrate and compounds of N, Mn, Fe and S, to transfer their electrons from respiration. In flooded systems, the decomposition of organic matter occurs almost entirely through the work of anaerobic and/or facultative microorganisms [152]. As anaerobic bacteria operate at a much lower energy level than aerobic ones, both decomposition and assimilation are slower in submerged systems. This fact is illustrated by the accumulation of plant residues in swamps and submerged sediments [153].

Humic acids (present in humic substances) have functional groups, such as carboxylic groups that can form metal salts, treatment with humic acids is likely to remove metals. Some of the metals can be associated with carbohydrates or amino acids. It is possible, however, that humic acid behaves like a macrocyclic compound for metals. These compounds have unusual ion-binding abilities. Its ability to form stable compounds with the alkali and alkali metal ions. In fact, alkaline and alkaline earth hydroxide solutions make up the bulk of the inorganic portion of humic acids. Part of heavy metals can also be associated with humic acids by the same mechanism that allows the complexation of metal ions by natural polyelectrolytes, such as polygalaturonic acid, or other polyacids. The addition of metals of such polymers forms insoluble complexes in the form of a "sandwich" [154].

7. Conclusions

Related evidence in the literature shows that attributes related to the growth and development of mangrove plants are negatively affected by their exposure to stress conditions by iron and nickel. However, organic matter, governed by its main component (such as humic substances) can, at least in part, assist recovery due to the damage caused by the aforementioned trace elements, showing the potential of humid organic matter to recover degraded environments and/or anthropized, such as the various mangroves located around the world.

Acknowledgements

Work funded by Fundação de Amparo à Pesquisa e Inovação do Espírito Santo (FAPES/2014-2018) and Conselho Nacional de Desenvolvimento Científico e Tecnológico (Process: 483518/2013-1).

Conflict of interest

The authors declare no conflict of interest.

Author details

Juliano de Oliveira Barbirato[1*], Natália C. Ferreira[2] and Leonardo B. Dobbss[3]

1 Instituto Estadual de Meio Ambiente – IEMA e Recursos Hídricos, Cariacica, Espírito Santo, Brasil

2 Universidade Vila Velha, Vila Velha, Espírito Santo, Brasil

3 Instituto de Ciências Agrárias da Universidade Federal dos Vales do Jequitinhonha e Mucuri, Unaí, Minas Gerais, Brasil

*Address all correspondence to: juliano.barbirato@outlook.com

References

[1] Francisco PM, Mori GM, Alvez FM, Tambarussi EV, Souza AP. Population genetic structure, introgression and hybridization in the genus Rhizophora along the Brazilian coast. Ecology and Evolution. 2018; 3491-3504. DOI: 10.1002/ece3.3900.

[2] Spalding MBF, Blasco F, FIELD C. World mangrove atlas. Okinawa: The International Society for Mangrove Ecosystems .ISME; 1997. 202 p.

[3] Schaeffer-Novelli Y. Manguezal ecossistema entre a terra e o mar. [S.l: s.n.]; 1995. 64p.

[4] Agraz-Hernández CM, Del Río-Rodríguez, RE, Chan-Keb CA, Saenz J, Muñiz-Salazar R. Nutrient removal efficiency of *Rhizophora mangle* (L.) seedlings exposed to experimental dumping of municipal waters. Diversity; 2018. Available from: www.mdpi.com/journal/diversity

[5] Odum WE, Mcivor CC, Smith TJ. Ecology of the mangroves of South Florida: a community profile. National Coastal Ecosystems Team, Fish and Wildlife Service. US Department of the Interior;1982. 144p

[6] Mallick I, Bhattacharyya C, Mukherji S, Dey D, Sakar SC, Mukhopadhyay UK, Ghosh A. Effective rhizoinoculation and biofilm formation by arsenic immobilizing halophilic plant growth promoting bacteria (PGPB) isolated from mangrove rhizosphere: A step towards arsenic rhizoremediation. Science of the Total Environment. 2018. 610(611):1239-1250. DOI: https://doi.org/10.1016/j.scitotenv.2017.07.234

[7] Sippo JZ, Lovelock CE, Santos IR, Sanders CJ, Maher DT. Mangrove mortality in a changing climate: An overview. Estuarine, Coastal and Shelf Science. 2018. 215:241-249. DOI: https://doi.org/10.1016/j.ecss.2018.10.011.

[8] Prefeitura de Vitória. MANGUEZAIS. 2018. Available from: https://m.vitoria.es.gov.br/cidade/meio-ambiente

[9] Ferreira RD. Os manguezais da baia de Vitória (ES): Um estudo de geografia física integrada. 1989 [Thesis]. Programa de pós-graduação em geografia, Faculdade de Filosofia, Letras e Ciências Humanas (USP), São Paulo, 1989.

[10] Campos Júnior CT. A construção da cidade: formas de produção imobiliária em Vitória. Flor & Cultura Editor; 2002. 161p.

[11] Bernini E, Rezende CE. Estrutura da vegetação em florestas de mangue do estuário do rio Paraíba do Sul, Estado do Rio de Janeiro, Brasil. Acta Bot. Brasil; 2004. 18(3):491-502. DOI: 10.1590/S0102-33062004000300009.

[12] Moscatelli M, Carli C. Avaliação preliminar do reflorestamento de manguezais: Lagoa Rodrigo de Freitas. In: Anais do III Simpósio de Ecossistemas da Costa Brasileira. Subsídios a um gerenciamento ambiental. São Paulo: ACIESP; 1994. p.131-134,

[13] Bernini E, Rezende CE. Variação estrutural em florestas de mangue do estuário do rio Itabapoana, ES-RJ. Biotemas. 2010;23(1):49-60. DOI: 10.5007/2175-7925.2010v23n1p49.

[14] Canellas LP, Moraes AA, Rumjanek VM, Garcia MR, Guridi F, Santos GA, Brazfilho R. Structural features of humic acids isolated from soils amended with urban organic resudues: na elemental composition, 13C NMR and Py-GC/MS analysis. Rev. Bras. Ci. Solo. 2002;26: 333-341.

[15] Dobbss LB, Canellas, LP, Alleoni LRF, Rezende CE, Fontes MPF, Velloso ACX. Eletroquímica

de latossolos brasileiros após a remoção da matéria orgânica humificada solúvel. R. Bras. Ci. Solo. 2008;32:985-996. DOI: https://doi.org/10.1590/S0100-06832008000300008

[16] Laanbroek HJ, Zhang QF, Leite M, Verhoeven JTA, Whigham DF. Effects of *Rhizophora mangle* leaf litter and seedlings on carbon and nitrogen cycling in salt marshes – potential consequences of climate-induced mangrove migration. Plant Soil. 2018;426: 383-400. DOI: 10.1007/s11104-018-3611-z.

[17] Atwood TB, Connolly RM, Almashasheer H, Carnell PE, Duarte CM, Lewis CJE, Irigoien X, Kelleway JJ, Lavery PS, Mecreadie PI, Serrano O, Sanders C, Santos I, Steven ADL, Lovelock CE. Global patterns in mangrove soil carbon stocks and losses. Nature Climate Change. Advance Online Publication. 2017. Available from: www.nature.com/natureclimatechange

[18] Clough B. Mangrove forest productivity and biomass accumulation in Hinchinbrook Channel, Australia. Mangroves and Salt Marshes 1998;2:191-198. DOI:10.1023/A:1009979610871.

[19] Dittmar T, lara, RJ. Do mangroves rather than rivers provide nutrients to coastal environments south of the Amazon River? Evidence from long-term flux measurements. Marine Ecology Progress Series. 2001;213:67-77. DOI: 10.3354/meps213067.

[20] Jennerjahn TC, Ittekkot V. Relevance of mangroves for the production and deposition of organic matter along tropical continental margins. Naturwissenschaften. 2002;89:23-30. DOI: 10.1007/s00114-001-0283-x.

[21] Krishna-Prasad MB, Ramanathan AL. "Organic Matter Characterization in a Tropical Estuarine-Mangrove Ecosystem of India: Preliminary Assessment by Using Stable Isotopes and Lignin Phenols," Estuar. Coast. Shelf Sci. 2009;84:617-624. DOI: 10.1016/j.ecss.2009.07.029

[22] Alongi DM. The energetics of mangrove forests. Dordrecht:Springer; 2009. p.216.

[23] Odum WE, Heald EJ. The detritus-based food web of an estuarine mangrove community. In: Cronin, L.E. (Ed.), Estuarine Research. Academic Press: New York; 1975. p. 265-286.

[24] Osland MJ, Feher LC, Griffith KT, Cavanaugh KC, Enwright NM, Day RH, Stagg CL, Krauss KW, Howard RJ, Grace JB, Rogers K. Climatic controls on the global distribution, abundance, and species richness of mangrove forests. Wiley; 2016. p.341-359.

[25] Rossi M, Mattos IFA. Solos de mangue do estado de São Paulo: caracterização química e física. Revista do Departamento de Geografia, 2002;15:101-113.

[26] Cintrón G, Schaeffer-Novelli Y. Caracteristicas y desarrollo estructural de los manglares de norte y sur América. Ciência Interamericana; 1985.25:4-15.

[27] Grasso M, Tognella MMP, Schaeffer-Novelli Yl. Aplicação de técnicas de avaliação econômica ao ecossistema manguezal. 1995.155p

[28] Coelho Junior C. Manguezal, Desenvolvimento estrutural da cobertura vegetal ao longo de gradientes de inundação, município de Cananéia, São Paulo, Brasil. São Paulo, 108p. [thesis]– Instituto Oceanográfico, Universidade de São Paulo; 1998.

[29] Novelli YS, Lacerda LD. Lagoas costeiras, manguezais, marismas, dunas e restingas. In: PANITZ, C.M.N.(Coord.).Diagnóstico ambiental oceânico e costeiro das regiões Sul e

Sudeste do Brasil. São Paulo: Fundespa; Iousp; Petrobrás; 1994. p.128-347.

[30] Chapman VJ. Mangrove biogeography. In: International Symposium on Biology and Management of Mangroves, Honolulu, Honolulu: East-West Center, 1975. p.3-22. 1975.

[31] Ab'Saber AN. Contribuição à geomorfologia do litoral paulista. Revista Brasileira de Geografia. 1955;17:1-48.

[32] Rizzini CT. Tratado de fitogeografia do Brasil. Aspectos ecológicos. Hucitec / Edusp, São Paulo. 1979. 747p.

[33] Tricart J. Ecodinâmica. Fundação Instituto Brasileiro de Geografia e Estatística. Superintendência de Recursos Naturais e Meio ambiente. Rio de Janeiro. 1977. 97p.

[34] Odum EP. Ecologia. Trad CG. Ottenwaelder, 3.ed. México, Nueva Editorial Interamericana. 1972.639 p.

[35] Cintrón G, Schaeffer-Novelli Y. Introducción a la ecologia del manglar. Montevideo, UNESCO-ROSTLAC. 1983.109p.

[36] Zhang X, Treitz PM, Chen D, Quan, C, Shi L. Li X. Mapping mangrove forests using multi-tidal remotely-sensed data and a decision-tree-based procedure. Int J Appl Earth Obs Geoinformation.2017;62:201-214. DOI: 10.1016/j.jag.2017.06.010.

[37] Laegdsgaard P, Johnson C. Why do juvenile fish utilise mangrove habitats? J. Exp. Mar. Biol. Ecol. 2001;257:229-253. DOI: 10.1016/S0022-0981(00)00331-2.

[38] Andreote FD, Jiménez DJ, Chaves D, Dias ACF, Luvizotto DM. The Microbiome of Brazilian Mangrove Sediments as Revealed by Metagenomics. PLoS ONE. 2012;6. DOI: 10.1371/journal.pone.0038600.

[39] Giri C, Ochieng E, Tieszen LL, Zhu, Z, Singh A, Loveland T, Masek J, Duke N. Status and distribution of mangrove forests of the world using earth observation satellite data. Glob. Ecol. Biogeogr. 2011;20:154-159. DOI: 10.1111/j.1466-8238.2010.00584.x.

[40] Sahoo K, Dhal N. Potential microbial diversity in mangrove ecosystems: A review. IJMS, 2009;38(2):249-256.

[41] Schaeffer-Novelli Y, Cintrón-Molero G, Adaime RR. Variability of mangrove ecosystems along the Brazilian coast. Estuaries. 1990. p.204-218.

[42] Vale CC, Ferreira RD. Os manguezais do litoral do Estado do Espírito Santo. In: Anais do Simpósio de Ecossistemas da Costa Brasileira. São Paulo: ACIESP; 1998. p. 88-94.

[43] Carmo TMS, Almeida R, Oliveira AR, Zanotti-Xavier S. Caracterização de um trecho do manguezal do rio da Passagem, baía de Vitória, Vitória-ES: In Anais do IV Simpósio de Ecossistemas Brasileiros (S. Watanabe, coord.). Águas de Lindóia: ACIESP; 1998. p.6-19.

[44] Almeida R. Conservação dos manguezais capixabas e sua importância para a diversidade biológica. p. 61-67. In: Menezes, L.F.T; Pires, F.R.; Pereira, O.J. (Orgs.) Ecossistemas costeiros do Espírito Santo: conservação e preservação. Vitória, EDUFES; 2007. 298p.

[45] Xu S, He Z, Zhang Z, Guo Z, Guo W, Lyu H, Li J, Du Z, Huang Y, Zhou R, Zhong C, Boufford DE, Lerdau M, Wu C, Duke NC. The International Mangrove Consortium; In: Shi S. The origin, diversification and adaptation of a major mangrove clade (Rhizophoreae) revealed by whole-genome sequencing. National Science Review, 2017;4:721-734. 2017.

[46] Tomlinson PB. The botany of mangroves. Cambridge University Press, Cambridge,United Kingdom. 1986.

[47] Menezes NL. Rhizophores in *Rhizophora mangle* L.: an alternative interpretation of socalled "aerial roots". Anais da Academia Brasileira de Ciências 2006;78: 213-226. DOI: 10.1590/S0001-37652006000200003.

[48] Prance GT, Silva MF, Albuquerque B, Araújo IJF, Carreira LM, Braga MM, Macedo M, Conceição PN, Lisboa PL, Braga PI, Lisboa RL, Vilhena RQ. Revisão taxonômica das espécies amazônicas de Rhizophoraceae. Acta Amazonica 1975;5(1):5-22. DOI: https://doi.org/10.1590/1809-43921975051005.

[49] Pascoalini SS, Lopes DMS, Falqueto AR, Tognella MMP. Abordagem ecofisiológica dos manguezais: uma revisão. Revista Biotemas. 2014;27(3):01-11.

[50] Charparzadeh N, Lucia M, Negad R, Izzo R, Izzo F. Antioxidative responses of *Calendula officinalis* under salinity conditions. Plant Physiol Biochem. 2004;42:695-701. DOI: https://doi.org/10.5007/2175-7925.2014v27n3p1

[51] Ball MC. Photosynthesis in mangroves. Wetlands: madison, 1986;6: 12-22.

[52] Sobrado MA. Leaf characteristics and gas exchange of the mangrove *Laguncularia racemosa* as affected by salinity. Photosynthetica. 2005;43(2):217-22.DOI: 10.1007/s11099-005-0036-8.

[53] Naidoo G, Tuffers AV, Von Willert DJ. Changes in gas exchange and chlorophyll fluorescence characteristcs of two mangrove and mangrove associate in response to salinity in the natural environment. Trees-Structure and function. 2002;16:140-146. DOI: 10.1007/s00468-001-0134-6.

[54] Antelmo L. Pigmentos fotossintéticos em *Rhizophora mangle* L. e *Laguncularia racemosa* gaerth. como indicadores das condições ambientais no manguezal da ilha de Vitória-ES. [thesis]. Universidade Federal do Espírito Santo. 2013.

[55] Bron IU, Ribeiro RV, Azzolini M, Jacomino AP, Machado EC. Chlorophyll fluorescence as a tool to evaluate the ripening of "Golden" papaya fruit. Postharvest Biology and Technology. 2004;33:163-173. DOI: 10.1016/j.postharvbio.2004.02.004.

[56] Arnon DI, Stout PR. The essentiality of certain elements in minute quantity for plants with special reference to copper. Plant physiol: Washington. 1939;14:371-375. DOI: https://dx.doi.org/10.1104%2Fpp.14.2.371.

[57] Cunha ACM, Paiva, HN, Xavier, A, Otoni, WC. Papel da Nutrição Minral na formação de raízes adventícias em plantas lenhosas. Pesquisa Florestal Brasileira. 2009;58: 35-47. DOI: 10.4336/2009.pfb.58.35.

[58] Prado RM. Efeito da aplicação de calcário no desenvolvimento, estado nutricional e produção de frutos de goiabeira e de caramboleira durante três anos em pomares em implantação. Jaboticabal [thesis]. Universidade Estadual Paulista, Faculdade de Ciências Agrárias e Veterinárias. 2003, 68p.

[59] Taiz L, Zeiger E. Fisiologia vegetal. Porto Alegre: Artmed. 5th ed. 2013. 918p.

[60] Freitas FA, Kopp MM, Sousa RO, Zimmer PD, Carvalho FIF, Oliveira AC. Absorção de P, Mg, Ca e K e tolerância de genótipos de arroz submetidos a estresse por alumínio em sistemas hidropônicos. Cienc. Rural. 2006;36(01):72-79. DOI: https://doi.org/10.1590/S0103-84782006000100011.

[61] Lynch, JP, St Clair SB. Mineral stress: the missing link in understanding how global climate change will affect plants in real world soilsField Crops Research. 2004;90:101-115. DOI: https://doi.org/10.1016/j.fcr.2004.07.008.

[62] Martínez-Ballesta MC, Dominguez-Perles R, Moreno DA, Muries B, Alcaraz-Lopez C, Bastias E, Garcia-Viguera C, Carvajal M. Minerals in plant food: effect of agricultural practices and role in human health. A review. Agron. Sustain. Dev. 2010;230:295-309. DOI: 10.1051/agro/2009022.

[63] Flowers TJ, Colmer, TD. Salinity tolerance in halophytes. New Phytol. 2008;179(4):945-63. doi: 10.1111/j.1469-8137.2008.02531.x.

[64] Coutinho CFB, Tanimoto ST, Galli A, Garbellini GS, Takayama M, Amaral RB, Mazo LH, Avaca LA, Machado SAS. Pesticidas: mecanismo de ação, degradação e toxidez. Pesticidas: Revista de Ecotoxicologia e Meio Ambiente. 2005;15:65-72.

[65] Macfarlane GR, Burchett MD. Photosynthetic pigments and peroxidase activity as indicators of heavy metal stress in the grey mangrove, *Avicennia marina* (Forsk.) Vierh. Marine Pollution Bulletin; 2001;42(3):233-240. DOI: 10.1016/S0025-326X(00)00147-8.

[66] Ghnaya T, Zaier H, Baioui R, Sghaier S, Lucchini G, Sacchi GA., Lutts S, Abdelly C. Implication of organic acids in the long-distance transport and the accumulation of lead in *Sesuvium portulacastrum* and *Brassica juncea*, Chemosphere; 2013;90:1449-1454. DOI: 10.1016/j.chemosphere.2012.08.061.

[67] Amari T, Ghnaya T, Debez A, Taamali M, Youssef NB, Lucchini G, Sacchi GA, Abdelly C. Comparative Ni tolerance and accumulation potentials between *Mesembryanthemum crystallinum* (halophyte) and *Brassica juncea*: metal accumulation, nutrient status and photosynthetic activity, Journal of Plant Physiology. 2014;171(17):1634-44. doi: 10.1016/j.jplph.2014.06.020.

[68] Larcher W. Ecofisiologia vegetal. 1st ed. São Carlos: Rima; 2006. 531 p.

[69] Lichtenthaler HK, Rinderle U. The role of chlorophyll fluorescence in the detection of stress conditions in plants. CRC Crit Rev Anal Chem; 1988;19[Suppl]:S29-S85.

[70] Prasad MB, Ramanathan AL. Organic matter characterization in a tropical estuarine-mangrove ecosystem of India: Preliminary assessment by using stable isotopes and lignin phenols. Estuarine, Coastal and Shelf Science; 2009;84:617-624.

[71] Marschner H. Mineral nutrition of higher plants, London, Academic Press. 1986. p. 483-643.

[72] Alhdad GM, Zoerb C, Al-Azzawi MJ., Flowers TJ. Is the reduced growth of the halophyte *Suaeda maritima* under hypoxia due to toxicity of iron or manganese?Environmental and Experimental Botany. 2015;116:61-71. https://doi.org/10.1016/j.envexpbot.2015.03.002

[73] Guerinot ML, Yi Y. Iron: nutritious, noxious, and not readily available. Plant Physiology; 1994;104(3):815-820. https://dx.doi.org/10.1104%2Fpp.104.3.815.

[74] Sahrawat KL. Elemental composition of rice plant as affected by iron toxicity under field conditions. Commun. Soil Sci. Plant Anal; 2004;31:2819-2827. https://doi.org/10.1080/00103620009370630

[75] Becana M, Moran JF, Iturbe-Ormaetxe I. Iron-dependent oxygen free radical generation in plants subjected to environmental stress: toxicity and antioxidant protection. Plant and Soil. 1998;201:137-147. https://doi.org/10.1023/A:1004375732137

[76] Halliwell B, Gutteridge JMC. Role of free radicals and catalytic metal ions in human disease: an overview. Methods in Enzymology; 1990;186:1-85. https://doi.org/10.1016/0076-6879(90)86093-b

[77] Majerus V, Bertin P, Lutts S. Effects of iron toxicity on osmotic potential, osmolytes and polyamines concentrations in the African rice (*Oryza glaberrima* Steud.). Plant Science. 2007;173(2):96-105. https://www.researchgate.net/deref/http%3A%2F%2Fdx.doi.org%2F10.1016%2Fj.plantsci.2007.04.003.

[78] Witte CP, Tiller S, Isidore E, Davies HV, Taylor MA. Analysis of two alleles of the urease gene from potato: polymorphisms, expression, and extensive alternative splicing of the corresponding mRNA. Journal of Experimental Botany, Oxford. 2005;56(409):91-99. https://doi.org/10.1093/jxb/eri014.

[79] Polacco JC. Nitrogen metabolism in soybean tissue culture. II: urea utilization and urease synthesis require Ni^{+2}. Plant Physiology, Lancaster. 1977;59(5):827-830. 59(5):827-30. doi: 10.1104/pp.59.5.827.

[80] Pell AN, Schofield P, Stone WC. Rates of digestion of feeds measured in vitro with computers. Proceedings Cornell Nutrition Conference For Feed Manufactures. Proceedings... Cornell University, Ithaca, N.Y. 1994;13(17):74-81.

[81] Silva GS, Do Nascimento AS, De Sousa ER, Marques EP, Brandão ALM, Corrêa LB, Silva GS. Distribution and Fractionation of Metals in Mangrove Sediment from the Tibiri River Estuary on Maranhão Island, Amazon Region. Revista Virtual de Química, 2014;6:323.

[82] Becana M, Moran J, Iturbe-Ormaetxe I. Iron-dependent oxygen free radical generation in plants subjected to environmental stress: toxicity and antioxidant protection. Plant and Soi, 1998;201:137-147. Doi: https://doi.org/10.1023/A:1004375732137

[83] Benckiser G, Santiago HU, Watanabe, N, Ottow, JCG. Effect of fertilization on exudation, dehydrogenase activity, iron-reducing populations and Fe + formation in the rhizosphere of rice (*Oryza sativa* L.) in relation to iron toxicity. Plant and Soil. 1984;79:305-316.

[84] Becker M, Asch F. Iron Toxicity in Rice—Conditions and Management Concepts. Journal of Plant Nutrition and Soil Science, 2005;168:558-573. Doi: http://dx.doi.org/10.1002/jpln.200520504

[85] Howeler RH. Iron-Induced Oranging Disease of Rice in Relation to Physico-Chemical Changes in a Flooded Oxisol. Soil Science Society of America Journal, 1973;37: 898-903. Doi: https://doi.org/10.2136/sssaj1973.03615995003700060030x

[86] O'Connor GA, Lindsay WL, Olsen SR. Diffusion of Iron and Iron Chelates in Soil. Soil Science Society of America Journal, 1971;35: 407-410. Doi: https://doi.org/10.2136/sssaj1971.03615995003500030024x

[87] Fageria NK, Baligar VC, Wright RJ. Iron nutrition of plants: an overview on the chemistry and physiology of its deficiency and toxicity. Pesquisa Agropecuária Brasileira. 1990;25(4):553-570.

[88] Hell R, Stephan UW. Iron uptake, trafficking and homeostasis in plants. Planta, 2003;216: 541-551..

[89] Alexandre JR, Oliveira MLF, Santos TC, Canton GC, Conceição JM, Eutropio FJ, Cruz ZMA, Dobbss LB, Ramos AC. Zinco e ferro: de micronutrientes a contaminantes do solo. Natureza on line 201; 10 (1): 23-28.

[90] Fang X, Zhang B, Thisse B, Bloom GS, Thisse C. Is Essential for Cell Proliferation and Motility During Zebrafish Embryonic Development. Cytoskeleton (Hoboken, N.J.). 2015;72(8):422-33.

[91] Wilson NC, Saintilan N. Reproduction of the mangrove species Rhizophora stylosa Griff. As its southern latitudinal limit. Aquatic Botany. 2018;151:30-37. https://doi.org/10.1016/j.aquabot.2018.07.009.

[92] Gillette B. Nickel named «Allergen of the Year». ACDS adds to list of substances warranting more attention. Dermat Times 2008;4:15-16.

[93] Gupta UC. Micronutrientes e elementos tóxicos em plantas e animais . In: Ferreira ME, Cruz MCP, Raij B, Abreu CA. eds. Micronutrientes e elementos tóxicos na agricultura. Jaboticabal, CNPq/FAPESP/POTAFOS, 2001;13-43.

[94] Salt DE, Kato N, Krämer U, Smith RD, Raskin I. The role of root exudates in Nickel hyperaccumulation and tolerance in accumulator and nonaccumulator species of Thlaspi. In: Terry N, Banuelos G (Eds.) Phytoremediation of contaminated soil and water. CRS Press LLC, London. 2000. p.189-200.

[95] Izosimova A. Modelling the Interaction between Calcium and Nickel in the Soil-Plant System. Bundesforschungsanstalt für Landwirtschaft (Braunschweig: FAL). 2005. p.91-99.

[96] Baccouch S, Chaoui EL, El Ferjani E. Nickel toxicity induces oxidative damage in *Zea mays* roots. Journal of Plant Nutrition, New York. 2001;24(7):1085-1097. https://doi.org/10.1081/PLN-100103805.

[97] Devi SR, Prasad MNV. Membrane lipid alterations in heavy metal exposed plants. In: Prasad MNV. (Ed.) Heavy metal stress in plants: from biomolecules to ecosystems, Narosa, Publishing House New Delhi, India, 2004; p.127-145. https://doi.org/10.1007/978-3-662-07745-0_5.

[98] Sheoran IS, Singal HR, Singh R. Effects of cadmium and nickel on photosynthesis and the 2 enzymes of the photosynthetic carbon reduction cycle in pigeon pea (*Cajanus cajan* L.) 3 Photosynthesis Res; 1990;23: 345-351. https://doi.org/10.1007/BF00034865.

[99] Gajewska E, Sklodowska M, Slaba M, Mazur J. Effect of nickel on antioxidative enzyme activities, proline and chlorophyll contents in wheat shoots. Biol Plant; 2006;50: 653-659. https://doi.org/10.1007/s10535-006-0102-5.

[100] Saito M, Arakaki R, Yamada A, Tsunematsu T, Kudo Y, Ishimaru N. Molecular Mechanisms of Nickel Allergy. Int J Mol Sci. 2016;17(2):202. Doi:10.3390/ijms17020202.

[101] Chakraborty P, Ramteke D, Chakraborty S. Geochemical partitioning of Cu and Ni in mangrove sediments: Relationships with their bioavailability . Marine Pollution Bulletin. 2015;93(1-2):194-201. https://doi.org/10.1016/j.marpolbul.2015.01.016

[102] Schickler H, Caspi H. Response of antioxidative enzymes to nickel and

cadmium stress in hyperaccumulator plants of the genus Alyssum. Physiologia Plantarum, Copenhagen. 1999;105:39-44.

[103] Molas J. Changes in morphological and anatomical structure of cabbage (*Brassica oleracea* L.) outer leaves and in ultrastructure of their chloroplasts caused by an in vitro excess of nickel. Photosynthetica, Prague, 1997;34:513-522.

[104] Witte CP, Tiller S, Isidore E, Davies HV. Taylor MA. Analysis of two alleles of the urease gene from potato: polymorphisms, expression, and extensive alternative splicing of the corresponding mRNA. Journal of Experimental Botany, Oxford, 2005;56:91-99.

[105] Chaoui A, Ferjani E. Effects of cadmium and copper on antioxidant capacities, lignification and auxin degradation in leaves of pea (*Pisum sativum* L.) seedlings. Comptes rendus biologies. 2005;328:23-31. Doi: 10.1016/j.crvi.2004.10.001.

[106] Gomes Junior RA. Resposta antioxidativa de células in vitro de café (*Coffea arabica*) submetidas aos metais pesados cádmio (Cd) e níquel (Ni). Piracicaba, 2006. 134 p. Thesis. Escola Superior de Agricultura Luiz de Queiroz.

[107] Kabata-Pendias A, Pendias H. Trace Elements in Soils and Plants. CRC Press, Boca Raton, Florida. 1984.

[108] Krupa Z, Siedlecka A, Maksymiec W, Baszynski YT. *In vivo* response of photosynthetic apparatus of *Phaseolus vulgaris* L. to nickel toxicity. Journal of Plant Physiology,1993;142:664-668.

[109] Berton RS, Valadares JMAS, Camargo OA, Bataglia OC. Peletização do lodo de esgoto e adição de CaCO3 na produção de matéria seca e absorção de Zn, Cu e Ni pelo milho em três latossolos. Revista Brasileira de Ciência do Solo, 1997;21:685-691.

[110] Neue HU, Gaunt JL, Wang ZP, Becker-Heidmann P, Quijano C. Carbon in tropical wetlands. Geoderma; 1997;79(1-4):163-185. https://doi.org/10.1016/S0016-7061(97)00041-4

[111] Correia MEF, Andrade AG. Formação de serapilheira e ciclagem de nutrientes. In: Santos, GA, Camargo FAO. Fundamentos da matéria orgânica do solo: ecossistemas tropicais e subtropicais. Porto Alegre: Genesis; 1999. p.197-225.

[112] Reef R, Feller IC, Lovelock CE. Nutrition of mangrove. Tree physiology; 2010;30:1148-1160, 2010. https://doi.org/10.1093/treephys/tpq048.

[113] Ferreira NC, Barbirato JO. Explorando o solo – Técnicas básicas para campo e laboratório. Novas Edições Acadêmicas, 2019.

[114] EMBRAPA – Empresa brasileira de pesquisa agropecuária. Centro Nacional de Pesquisa de Solos. Sistema Brasileiro de Classificação de Solos. 3rd ed. Rio de Janeiro, 2013. p. 306.

[115] Correia MD, Sovierzoski HH. Ecossistemas marinhos: recifes, praias e manguezais. Série Conversando sobre Ciências em Alagoas. Maceió: EDUFAL, 2005. p.55.

[116] Kristensen E, Bouillon S, Dittmar T, Marchand C. Organic carbon dynamics in mangrove ecosystems: a review. Aquatic Botany; 2008;89(2):201-219, 2008. https://doi.org/10.1016/j.aquabot.2007.12.005

[117] Breemen NV, Buurman P. Soil formation. Dordrecht, Kluwer, 1998. 376p.

[118] Ferreira TO. Processos pedogenéticos e biogeoquímica de Fe

e S em solos de manguezais. [thesis] Escola Superior de Agricultura — "Luiz de Queiroz", Universidade de São Paulo, Piracicaba, 2006. p. 141.

[119] Otero XL, Macías F. Variation with depth and season in metal sulfides in salt marsh soils. Biogeochemistry, 2002;61:247-268. doi: 10.1023/A:1020230213864

[120] Manini E, Fiordelmondo C, Gambi C, Pusceddu A, Danovaro R. Bentic microbial loop functioning in coastal lagoons: a comparative approach. Oceanologica Acta, 2003;26:27-38. https://doi.org/10.1016/S0399-1784(02)01227-6.

[121] Sherman RE, Fahey TJ, Howarth RW. Soil-plant interactions in a neotropical mangrove forest: iron, phosphorus and sulfur dynamics. Oecologia, 1998;115:553-563. Doi: 10.1007/s004420050553.

[122] Grimsditch G, Alder J, Nakamura T, Kenchington R, Tamelander J. The blue carbon special edition – Introduction and overview. Ocean and Coastal Management, 2012:1:1-4. http://dx.doi.org/10.1016/j.ocecoaman.2012.04.020.

[123] Rezende CE. Ecologia e Biogeoquímica de Manguezal. In: Pereira PC, Gomes, AS. (Org.). Biologia Marinha. Rio de Janeiro: Editora Interciência. 2009. p.361-383.

[124] Alongi DM. Coastal Ecossistem processes. Boca Raton: CRC Press, 1998. 419p.

[125] Yücel M, Konovalov SK, Moore TS, Janzen, CP, Luther, GW. Sulfur speciation in the upper black sea sediments. Chemical Geology, 2010;269:364-375. https://doi.org/10.1016/j.chemgeo.2009.10.010

[126] Holmer M. The effect of oxygen depletion on anaerobic organic matter degradation in marine sediments. Estuarine, Coastal and Shelf Science, 1999;48:383-390. https://doi.org/10.1006/ecss.1998.0424.

[127] Dumat C, Quenea K, Bermond A. Study of the trace metal ion influence on the turnover of soil organic matter in cultivated contaminated soils. Environmental Pollution, Amsterdam. 2006;142:521-529.

[128] Ribeiro-Filho MR, Siqueira JO, Curi N, Simão JBP. Fracionamento e biodisponibilidade de metais pesados em solo contaminado, incubado com materiais orgânicos e inorgânicos. R. Bras. Ci. Solo. 2001;25:495-507.

[129] Santos NM, Accioly AMA, Nascimento CWA, Santos JAG, Silva IR. Ácidos húmicos e carvão vegetal ativado como amenizantes em solo contaminado por chumbo. R. Bras. Ci. Solo. 2014;38:345-351.

[130] Canellas LP, Santos GA. Humosfera: tratado preliminar sobre a química das substâncias húmicas. Campos dos Goytacazes, 2005. 309 p.

[131] Ferreira NC, Barbirato JO, Dobbss LB. Análise ambiental do principal rio de abastecimento do estado do Espírito Santo, Sudeste do Brasil. Brazilian Journal Animal of and Environmental Research, 2020;3(4): 3154-3169. Doi: 10.34188/bjaerv3n4-035

[132] Guerra AJT. O papel da materia organica e dos agregados na erodibilidade dos solos. Anuário do Instituto de Geociências, 1990;13:43-52.

[133] Matilainen A, Vepsalainen M, Sillanpaa M. Natural organic matter removal by coagulation during drinking water treatment: a review. Adv. Colloid Interf. 2010;159:189-197. https://doi.org/10.1016/j.cis.2010.06.007.

[134] Kononova MM. Matéria orgânica del suelo: su natureza, propriedades y

método de investigación. Barcelona, Oikos-Tour, 1982. p.365.

[135] Adani F, Genevini P, Tambone F, Montoneri E. Compost effect on soil humic acid: a NMR study. Chemosphere. 2006;65:1414-1418.

[136] Stevenson JF. Humus chemistry: Genesis, composition, reactions. 2. ed. New York, John WWiley e Sons. 1994. p.496.

[137] Lu XQ, Hanna JV, Johnson WD. Source indicators of humic substances: an elemental composition, solid state 13C CP/MAS NMR and Py-GC/MS study. Appl. Geochem. 2000;15:1019-1033.

[138] He MC, Shi YH, Lin CY. Characterization of humic acids extracted from the sediments of the various rivers and lakes in China. J. Environ. Sci. 2008;20:1294-1299.

[139] Dobbss L, Medici L, Peres L, Pino-Nunes L, Rumjanek V, Façanha A, Canellas L. Changes in root development of Arabidopsis promoted by organic matter from oxisols. Annals of Applied Biology, 2007;151: 199-211. Doi: https://doi.org/10.1111/j.1744-7348.2007.00166.x

[140] Dobbss LB, Canellas LP, Olivares FL, Aguiar, Oliveira N, Peres LEP, Azevedo M, Spaccini R, Piccolo A, Façanha AR. Bioactivity of Chemically Transformed Humic Matter from Vermicompost on Plant Root Growth. Journal of Agricultural and Food Chemistry, 2010;58:3681-3688.

[141] Aguiar NO, Canellas LP, Dobbss LB, Zandonadi DB, Olivares FL, Façanha AR. Distribuição de massa molecular de ácidos húmicos e promoção do crescimento radicular. Rev. Bras. Ci. do Solo. 2009;33:1613-1623.

[142] Bezerra LJC. Caracterização dos Tabuleiros Pré- Litorâneos do Estado do Ceará / Luiz José Cruz Bezerra. 2009. *Thesis*. Universidade Federal do Ceará, Instituto de Ciências do Mar – LABOMAR, Fortaleza, 2009.

[143] Marchand C, Fernandez J, Moreton B, Landi L, Lallier-Verges E, Baltzer F. The partitioning of transitional metals (Fe, Mn, Ni, Cr) in mangrove sediments downstream of a ferralitized ultramafic watershed (New Caledonia). Chem. Geol. 2012;70-80.

[144] Noël V, Morin G, Juillot F, Marchand C, Brest J, Bargar JR, Muñoz M, Marakovic G, Ardo S, Brown Jr GE. Ni cycling in mangrove sediments from New Caledonia. Geochimica et Cosmochimica Acta. 2015;169:82-98.

[145] Swaine DJ, Mitchell RL. Trace-element distribution in soil profiles. J. Soil Sci. 1960;11(2):347-368.

[146] Senesi N, Dorazio VD, Miano TM. Adsorption mechanisms of s-triazine and bipyridylium herbicides on humic acids from hop field soils. Geoderma. 1995;66:273-283. https://doi.org/10.1016/0016-7061(94)00083-M.

[147] Moraes SL, Rezende MOO. Capillary electrophoresis (CE): a powerful tool to characterize humic acid (HA). J. Brazil Chem. Soc. 2008;19, 24-28. Doi: 10.1590/S0103-50532008000100005.

[148] Piccolo A. The supramolecular structure of humic substances: A novel understanding of humus chemistry and implications in soil science. Adv. Agron., 2002;75:57-133. DOI: 10.1016/S0065-2113(02)75003-7.

[149] Santos NM, Accioly AMA, Nascimento CWA, Santos JAG, Silva IR. Ácidos húmicos e carvão vegetal ativado como amenizantes em solo contaminado por chumbo. R. Bras. Ci. Solo, 2014;38:345-351. DOI: 10.1590/S0100-06832014000100035.

[150] Zhou J, Diao X, Wang T, Chen G, Lin Q,Yang X, Xu J. Phylogenetic diversity and antioxidant activities of culturable fungal endophytes associated with the mangrove species *Rhizophora stylosa* and *R. mucronata* in the South China Sea. PLOS ONE 2018;13(6):e0197359. https://doi.org/10.1371/journal.pone.0197359.

[151] Martins CA, Nogueira NO, Ribeiro PH, Rigo MM, Candido AO. A Dinâmica De Metais-Traço No Solo. R. Bras. Agrociência, Pelotas, 2011;17(3-4):383-391. http://dx.doi.org/10.5935/1984-6835.20110014

[152] Milori DMBP, Martin Neto L, Bayer C, Mielniczuck J, Bagnato VS. Humification degree os soil humic acids determined by fluorescence spectroscopy. Soil Sci. 2002;67:739-749. DOI: 10.1097/00010694-200211000-00004.

[153] Zhang X, Treitz PM, Chen D, Quan C, Shi L, Li X. Mapping mangrove forests using multi-tidal remotely-sensed data and a decision-tree-based procedure. Int J Appl Earth Obs Geoinformation, 2017;62:201-214. DOI: 10.1016/j.jag.2017.06.010.

[154] Nissenbaum A, Swaine DJ. Organic–metal interactions in recent sediments: the role of humic substances. Geochim. Cosmochim. Acta, 1976;40:809-816.

Chapter 8

Chemical Characteristics of Humic Substances in Nature

Claudio Fernando Mahler, Nicoly Dal Santo Svierzoski and Cassiano Augusto Rolim Bernardino

Abstract

Humic substances are the main constituents of natural organic matter, found in both aquatic and terrestrial environments. Humic substances are a complex, dispersed, and heterogeneous mixture of various organic compounds synthesized from organic matter residues, decomposed by microorganisms. Most scientists indicate that humic substances are as a supramolecular association of small heterogeneous molecules stabilized by weak intermolecular bonds. When these substances are presented in water intended for drinking or industrial use, it can have a significant impact on the treatability of this water and on the success of chemical disinfection processes, due to possible formation of organic compounds harmful to human health. Moreover, the humic substances can be used of several ways such as fertilizer to help in the development of plants, to improve soil erosion and to removal of organic compounds and metals from soils and waters. In addition, humic substances suggest an important role in mitigating areas degraded by the phytoremediation technique. The purpose of chapter is to provide an overview of humic substances and to discuss their concepts, chemical characteristics, ecological effects and technological applications for soils and aquatic systems.

Keywords: Humic substances, Humic acid, Fluvic acid, Humin, Organic matter, Soils, Aquatic system

1. Introduction

Humic substances are the main constituents of natural organic matter, found in both aquatic and terrestrial environments. Natural organic matter can be classified into aquagenic organic matter and paedogenic organic matter. Aquagenic organic matter occurring in ocean waters is formed by the excretion and decomposition of plankton and aquatic bacteria. Paedogenic organic matter is produced by the decomposition of land plants and microorganisms, including leached material from the soil anchored in the aquatic system. This type of organic matter is formed by the degradation of lignin, polysaccharides and proteins, which lead to the formation of organic compounds of carboxylic, phenolic, benzoic, and aliphatic compounds. In addition, humic substances are the main carbon reservoir in the biosphere and account for approximately 70 to 90% of soil organic matter [1].

According to the International Humic Substances Society (IHSS), humic substances are complex heterogeneous mixtures of polydisperse materials formed by a humification process, in which chemical and biochemical reactions occur during the decomposition of plant and microbial residues [2]. In [3] was also noted this new understanding, where humic substances are heterogeneous and relatively small molecular components of soil organic matter in supramolecular associations with a variety of organic compounds of biological origin and synthesized by abiotic and biotic reactions in soil. The first publications on humic substances were reported in 1786 with the extraction of humic acids from peat by Achard in Germany and by Vauquelin in 1797 with the extraction of plant residues. Between 1829 and 1837, studies to understand the origin and composition of humic substances began with the research conducted by Sprengel, reporting that more alkaline soils exhibited greater abundance of humic acids, making them more fertile [4]. In the 19th and 20th centuries, humic substances were often examined through the acid–base theory. In the mid-20th century studies were initiated in order to classify humic fractions. The author in [5] created the nomenclatures of fulvic acids, humic acids and humins. This classification is based on their solubility in aqueous medium. Then, fulvic acids are soluble at acidic or alkaline pH, humic acids are soluble at alkaline pH, and humins are insoluble at any pH [4, 6, 7]. Researchers have been discussing, in recent years, models in order to explain the chemical structure of humic substances. Currently, there are three widely discussed models - the macromolecular, the micellar, and the supramolecular structure models. The macromolecular model assumes that humic substances are a soil polymer and that humification is the process by which organic compounds from plants and animals are not fully oxidized to CO_2 and H_2O and accumulate in the soil as humus [8–10].

The micellar model states that humic substances consist of macro-structures of high molecular mass and that the macromolecular properties result from associations of small molecular species in micellar structures [11]. The author in [12] defines humic substances as a supramolecular association of small heterogeneous molecules stabilized by weak intermolecular bonds, which can be broken by small amounts of organic acids. Among the three main structural models (micellar, macromolecular and supramolecular) of humic substances, currently, the supramolecular model is the most widely accepted by researchers in the IHSS [2].

However, the scientific community has been discussing the macromolecular and supramolecular models, due to the undefined chemical structure of humic substances.

2. Concept and characterization of humic substances

Researchers in this area has been discussing the concept, origin, and composition of humic substances for the last decades, and these questions have not yet been clarified. According to [13, 14] humic substances are a complex, dispersed, and heterogeneous mixture of various organic compounds synthesized from organic matter residues, decomposed by microorganisms.

Humic substances exhibit a wide variety of structures and chemical compositions. As an example, a considerable part of these substances present the benzene ring (phenols and quinones), as shown in **Figure 1**. In addition, several functional groups are present in humic substances such as carboxylic, hydroxyl (phenolic and alcoholic), carbonyl, and amino groups [15]. The authors [16] list the substances identified in humic extracts, that include mono-, di- and trihydroxy acids, fatty acids, dicarboxylic acids, linear alcohols, phenolic acids, terpenoids, steroid compounds, carbohydrates and amino acids.

In [9], the author reports that the functional groups that contribute most to the surface charge and reactivity of humic substances are phenolic and carboxylic groups. Humic acids behave as mixtures of dibasic acids, with pK_a value around 4

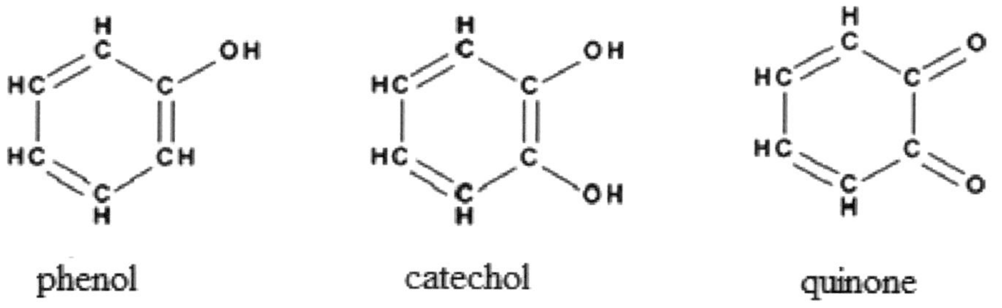

Figure 1.
Molecular building blocks that form humic substances containing the benzene ring.

for protonation of carboxylate groups and around 8 for protonation of phenolate groups. There is considerable overall similarity between the individual humic acids [17].

For this reason, the pK_a values measured for a given sample are mean values relative to the constituent species. The other important characteristic is charge density. Fulvic acids are defined as associations of small hydrophilic molecules which several acidic functional groups that form the fulvic clusters dispersed in solution at any pH. Humic acids are made by associations mainly of hydrophobic compounds which it are stabilized at neutral pH by hydrophobic dispersive forces such Van der Waals forces, π-π bonds, CH-π bonds. Their conformations grow progressively in size when intermolecular hydrogen bondings are increasingly, flocculate at lower pH [11]. The presence of carboxylate and phenolate groups gives humic acids the ability to form complexes with ions such as Mg^{2+}, Ca^{2+}, Fe^{2+} and Fe^{3+}. Many humic acids have two or more of these groups arranged in such a way that allows the formation of chelate complexes [18]. The development of complexes (chelates) is an important aspect of the biological role of humic acids in regulating the bioavailability of metallic ions [17].

According to [19] the elemental composition of humic substances can be divided and vary as follows: fulvic acids (35.1 to 75.7% carbon, 16.9 to 55.8% oxygen, 0.4 to 7.9% hydrogen, 0.5 to 8.2% nitrogen and 0.1 to 3.6% sulfur), humic acids (37.2 to 75.8% carbon, 7.9 to 56.6% oxygen, 1.6 to 11.7% hydrogen, 0.5 to 10.5% nitrogen and 0.1 to 8.3% sulfur) and humin (48.3 to 61.6% carbon, 28.8 to 45.1% oxygen, 7.3 to 14.2% hydrogen, 2.9 to 6.0% nitrogen and 0.1 to 0.9% sulfur).

The presence and relative abundance of fulvic acids, humic acids, and humin is inferred by laboratory extraction, a process that modifies their original state. Humic and fulvic acids are extracted as a colloidal solution from the soil to other solid phase sources in a strongly basic aqueous solution of sodium or potassium hydroxide. Humic acids are precipitated into the solution by adjusting the pH to 1 with hydrochloric acid, keeping fulvic acids in solution. This is the operational difference between humic and fulvic acids. Humin is insoluble in diluted alkali. The alcohol-soluble portion of the humic fraction is generally referred to as humic acid. The named "gray humic acids" (GHA), according to [20], are soluble in alkaline media of low ionic strength; "brown humic acids" (BHA) are soluble under alkaline conditions regardless of ionic strength; and fulvic acids (FA) are soluble regardless of pH and ionic strength. Humic acid, as traditionally produced in the laboratory, is not a single acid; instead, it is a complex mixture of many different acids containing carboxylic and phenolic groups so that the mixture behaves functionally as a dibasic acid or, occasionally, as a tribasic acid. Humic acid used for soil correction is manufactured using the same well-established procedures. Humic acids can form complexes with ions commonly found in the environment creating humic colloids. Humic acids are insoluble in water at acid pH, while fulvic acids are also derived from humic substances, but are water soluble throughout the pH range [21]. Humic and fulvic acids are often used as a soil supplement in agriculture and, less commonly, as a human nutritional supplement. As a nutritional supplement, fulvic acid can be found in liquid form as a component of mineral colloids. Fulvic acids are polyelectrolytes and are unique colloids that diffuse easily through membranes, whereas all other colloids do not [22]. A sequential chemical fractionation called Humeomics can be used to isolate more homogeneous humic fractions and determine their molecular structures by advanced spectroscopic and chromatographic methods [23].

3. Determination of humic substances in water samples

The water naturally found in springs has organic matter and microorganisms that can be pathogenic. The organic matter originates from the remains of dead plants and animals and is carried to the springs by surface runoff after rainfall. Domestic and industrial effluents discharged directly into water bodies also contribute to an increase in the amount of organic matter in the springs. The presence of humic substances in water intended for drinking or industrial use can have a significant impact on the treatability of this water and on the success of chemical disinfection processes. For example, according to [24, 25] humic and fulvic acids can react with the chemicals used in the chlorination process, forming disinfection by-products, such as dichlorocetonitriles, which are toxic to humans. In addition, the color occurring in natural waters is largely due to the presence of humic substances. When found in water distributed for human consumption, they impart a dark color and taste to the water. Moreover, they can form by-products during pre-oxidation and disinfection and can complex with metals and organic micropollutants. Such by-products, if originating from the reaction with chlorine, are potentially carcinogenic chlorinated organic compounds, known as trihalomethanes (THM) [26, 27]. Therefore, accurate methods for establishing humic acid concentrations are essential for maintaining water supplies, especially from watersheds. Since many different bio-organic molecules, in many diverse physical associations, are mixed together in natural environments, it is complicated to measure their precise concentrations in the humic superstructure. For this reason, humic acid concentrations are traditionally estimated from organic matter concentrations (usually from concentrations of total organic carbon or dissolved organic carbon). The organic carbon dissolved in waters accounts for 1/3 to 1/2 of the total in natural waters [28]. Humic substances, mainly humic acid, represent the largest fraction of dissolved organic carbon present in surface water and groundwater [29]. Humic extracts are composed of a large number of different bio-organic molecules that have not yet been fully separated and identified. However, unique classes of residual biomolecules have been identified by selective extractions and chemical fractionation and are represented by alkanoic and hydroxy-alkanoic acids, resins, waxes, lignin residues, sugars, and peptides. **Table 1** shows several methods that are used in the extraction of humic substances, such as lyophilization, co-precipitation, ultrafiltration, solvent extraction and resins [30, 31]. The separation of humic substance in water by adsorption chromatography has been widely employed [14, 32, 33].

The authors in [34] developed a methodology based on resin adsorption and carbon concentrations. The method is quantitative for determining the levels of humic substances in water. According to [35], this method was evaluated as the most appropriate and it produces more accurate results due to the use of spectrophotometry to evaluate only part of the humic substances present in a given sample.

Method	Benefit	Limitation
Lyophilization	Soft method and high concentration factor	Solutes are concentrated, less the volatile
Co-precipitation	Good cost-effective and effective for waters with a high content of dissolved organic carbon.	Low efficiency for large sample volumes.
Ultrafiltration	Fractionation of solutes by molecular weight and indicated for large sample volumes	Interaction with the membrane and clogging may occur
Solvent extraction	Exclusion of inorganic salts	Slow method and irreversible interactions occur in the sample/solvent
Resins	Suitable for large sample volumes, high concentration factor and easy regeneration of adsorbent	Changes in the sample and contamination from the resin

Table 1.
Benefits and limitations of the applied methods for extracting humic substances in aquatic environment.

Another technique used to determine the concentration of humic substances in water is using visible and ultraviolet spectroscopy. For this purpose, a standard absorbance vs. concentration line is constructed and thus it is possible to determine the concentration of humic substances based on the absorbance readings of the samples (aqueous solution of humic substances). The authors in [31] performed the isolation of humic substances from water and used this technique to determine the content of aromatic compounds and the color intensity of humic substances as well as the analysis of dissolved organic carbon for determine the total concentration of humic substances. Others authors, as cited in [36, 37], also used these techniques in their studies of humic substances.

4. Ecological effect

The additives of soil organic matter have been known by farmers to be beneficial to plant growth for longer than recorded history. However, the chemistry and function of organic matter have been the subject of controversy since humans began to postulate about it in the 18th century. Until Liebig's time, it was assumed that humus was used directly by plants but after Liebig showed that plant growth depends on inorganic compounds, many soil scientists argued that organic matter was useful for fertility only when it was decomposed with the release of its constituent nutrient elements in inorganic forms. Currently, soil scientists take a more holistic view and have recognized that humus influences soil fertility through its effect on water holding capacity of soil. Moreover, since it has been shown that plants absorb and translocate the complex organic molecules of systemic insecticides, they can no longer discredit the idea that plants may be able to absorb the soluble forms of humus.

The positive effects of humic substances on plant growth may be related to indirect effects, such as increased efficiency in fertilization and reduced soil compaction, or to direct effects, such as improvements in plant biomass [38, 39]. Such substances induce growth and increase the absorption of nitrogen, phosphorus and potassium [40]. Nitrogen is responsible for the formation of amino acids, proteins, enzymes and nucleotides, which ensures better growth and functioning of plants. Humic substances can also improve plant metabolism and absorption. In summary, the different fractions of humic substances can affect plant growth and development, leading to structural and physiological improvements. In addition, it favors the development of rhizoids (specialized epidermal cells whose function is water absorption and anchoring) and lateral roots [41]. The authors in [42] reported that humic substances induce the presence of nitric oxide in the lateral roots, which promotes their development. Root exudates also benefit from humic substances, which are responsible for the solubilization and mobility of nutrients. In addition, the use of humic substances in the soil is possible to observe the length of the rhizoid, the density and cell proliferation in the root [41]. Another important factor of these substances in the soil is the ability to improve the soil's retention capacity, decreasing erosion processes and thus, favoring a good soil structure which influences its workability, nutrient availability and development and growth of agricultural crops [9, 11].

Finally, in aquatic environments, humic substances are naturally present in dissolved, particulate or suspended forms and their importance is related to the availability of organic and inorganic nutrients for bacteria, fungi, phytoplankton and aquatic macrophytes. Another relevant factor is the relationship between humic substances with the complexation, absorption and immobilization of organic contaminants and heavy metals, which consequently increases the bioavailability and availability for organisms [1, 43].

5. Technological applications

Scientific research has shown different technological applications related to humic substances to improve the quality of water and soil.

In the aquatic environment, the humic acid binding skills of heavy metals have been exploited to develop remediation technologies for the removal of heavy metals from wastewater. For this, the authors in [44] used nano magnetic particles coated with humic acids. After capturing the lead ions, the nanoparticles can be captured using a magnet. On the other hand, to guarantee the potability of water for human consumption, it is necessary to remove humic substances due to the presence of organic compounds harmful to human health. Then, in the last few decades, several methods to improve water quality can be applied such as treatment methods, including coagulation, filtration and advanced oxidation processes [45]. The use of heterogeneous photocatalysis using titanium dioxide was analyzed by [46] to degrade the organic matter in water for human consumption. The results showed the complete removal of humic substances from the clarifier was after 220 minutes of irradiation.

Another important issue is the landfill leachate which is rich in humic substances, such as humic acid and fulvic acid, and these substances are important constituents of organic fertilizers used for plant growth. A study that evaluated the removal of humic substances from leachate for use as fertilizer was presented by [47]. The authors concluded that when using 100 ppm of the concentrate of humic substances extracted from the leachate, the green bean seedlings had an increase of 54.7% in height and 121.4% in weight, verifying that there was no phytotoxicity of the extracted.

Regarding the application of substances to the soil, the authors in [48] studied the effects of humic acid on plant growth, observing that "humic acids accelerate the growth of plants" and there are "relatively large responses to low application speeds". In [49] the authors showed that the addition of humate to the soil significantly increases the mass of the roots in creeping grass of folded grass. In the field of soil mechanics and geology, the term humate refers to geological materials, such as coal beds, mudrock or pore material in sandstones, which are rich in humic acids. The authors in [50] report that humate has been extracted from the Fruitland Formation of New Mexico for use as a soil corrector since 1970, with almost 60,000 metric tons produced in 2016. According to [51] humate deposits may also play an important role in the genesis of uranium ore bodies. Another contribution of humic substances refers to their potential use in reducing soil erodibility [52–54]. The authors in [55] evaluated the incorporation of humic substances extracted from the coal tailings in the material of the gneiss mining. The authors concluded that the use of these substances favored the reduction of soil erodibility in mining areas. This way of applying humic substances allows an alternative use to the tailings to be implanted and still providing the fight against erosion of degraded areas.

In the case of heavy metals, the authors in [56] evaluated the ability to simultaneously remove a soil artificially contaminated with copper, lead, zinc, cadmium and chromium through the use of humic substances extracted from the compost of livestock manure. The results showed that washing the soil with the studied compound allowed a greater capacity to remove metals from the soil due to the variety of structures found in the material used. In addition, the authors concluded that the metals have become more bioavailable for plants. Therefore, the use of humic substances combined with the phytoremediation technique is an excellent strategy to mitigate areas contaminated with heavy metals [57]. This technique uses plants to reduce the levels of contaminants to levels safe and compatible with the preservation of human health, preventing or hindering the spread of substances harmful to the environment [58]. Studies report that different species of plants are highly efficient for mitigating metals in soils when these elements are bioavailable [59]. Therefore, the use of humic substances in soils is desirable as it improves development and metals are more bioavailable for plants in the phytoremediation process.

6. Final considerations

Humic substances have made great strides in understanding the supramolecular structure and it have been widely accepted by several researchers. However, its structure is not yet fully defined and several issues remain under discussion by the scientific community. The fact is that these substances can be used to improve the quality and structure of the soil. Humic substances have unique chemical properties that benefit the soil in different ways, for example, it acts to combat soil erosion, improves the development of plants and removes pollutants from the soil. These characteristics indicate that humic substances play an important role in mitigating areas degraded by the phytoremediation technique. As for the aquatic system, humic substances can be used in the remediation of wastewater. In the treatment of leachate in landfills, humic substances can act as organic fertilizers for agricultural soils. On the other hand, these substances are a problem in drinking water for human consumption and must be treated efficiently to avoid damage to human health.

Author details

Claudio Fernando Mahler*, Nicoly Dal Santo Svierzoski
and Cassiano Augusto Rolim Bernardino
Department of Civil Engineering, COPPE, Federal University of Rio de Janeiro,
Rio de Janeiro, Brazil

*Address all correspondence to: cfmahler@gmail.com

References

[1] Lipczynska-Kochany, E., 2018a. Humic substances, their microbial interactions and effects on biological transformations of organic pollutants in water and soil: A review. Chemosphere 202, 420-437. https://doi.org/10.1016/j.chemosphere.2018.03.104

[2] IHSS, 2021. What are humic substances? [WWW Document]. Internacional Humic Substances Society. URL http://humic-substances.org/ (accessed 3.15.21).

[3] Piccolo, A., Spaccini, R., Drosos, M., Vinci, G., Cozzolino, V., 2018. The molecular composition of humus carbon: Recalcitrance and reactivity in soils, in: The Future of Soil Carbon. Elsevier, pp. 87-124. https://doi.org/10.1016/B978-0-12-811687-6.00004-3

[4] Santos, L.L. dos, Lacerda, J.J.J., Zinn, Y.L., 2013. Partição de substâncias húmicas em solos brasileiros. Revista Brasileira de Ciência do Solo 37, 955-968. https://doi.org/10.1590/S0100-06832013000400013

[5] Odén, S. Die Huminsäuren. Kolloidchem Beih 11, 75 (1919). https://doi.org/10.1007/BF02557436

[6] Hayes, M.H.B., 1998. Humic substances: Progress towards more realistic concepts of structures, in: Humic Substances. Elsevier, pp. 1-27. https://doi.org/10.1016/B978-1-85573-806-5.50005-8

[7] Zech, W., Senesi, N., Guggenberger, G., Kaiser, K., Lehmann, J., Miano, T.M., Miltner, A., Schroth, G., 1997. Factors controlling humification and mineralization of soil organic matter in the tropics. Geoderma 79, 117-161. https://doi.org/10.1016/S0016-7061(97)00040-2

[8] Kononova, M.M., 1982. Matéria orgânica del suelo: su naturaleza, propriedades y métodos de investigación. Oikos-Tou, Barcelona.

[9] Schnitzer, M., Khan, S.U., 1972. Humic substances in the environment. Marcel Dekker, New York.

[10] Stevenson, J.F., 1994. Humus chemistry: genesis, composition, reactions. John Wiley e Sons INC, New York.

[11] Wershaw, R., 1993. Model for humus in soils and sediments. Environmental Science & Technology 27, 814-816. https://doi.org/10.1021/es00042a603

[12] Piccolo, A., 2002. The supramolecular structure of humic substances: A novel understanding of humus chemistry and implications in soil science, in: Advances in Agronomy. pp. 57-134. https://doi.org/10.1016/S0065-2113(02)75003-7

[13] Tissot, B.P., Welte, D.H., 1984. Petroleum formation and occurrence. Springer Berlin Heidelberg, Berlin, Heidelberg. https://doi.org/10.1007/978-3-642-87813-8

[14] Thurman, E.M., 1985. Aquatic Humic Substances, in: Organic Geochemistry of Natural Waters. Springer Netherlands, Dordrecht, pp. 273-361. https://doi.org/10.1007/978-94-009-5095-5_11

[15] Libes, S.M., 2009. Introduction to Marine Biogeochemistry, 2nd ed. Elsevier.

[16] Drosos, M., Piccolo, A., 2018. The molecular dynamics of soil humus as a function of tillage. Land Degradation & Development 29, 1792-1805. https://doi.org/10.1002/ldr.2989

[17] Ghabbour, E.A., Davies, G., 2001. Humic substances: Structures, models and functions. Royal Society of

Chemistry, Cambridge. https://doi.org/10.1039/9781847551085

[18] Tipping, E., 1994. WHAMC—A chemical equilibrium model and computer code for waters, sediments, and soils incorporating a discrete site/electrostatic model of ion-binding by humic substances. Computers & Geosciences 20, 973-1023. https://doi.org/10.1016/0098-3004(94)90038-8

[19] Rice, J.A., MacCarthy, P., 1991. Statistical evaluation of the elemental composition of humic substances. Organic Geochemistry 17, 635-648. https://doi.org/10.1016/0146-6380(91)90006-6

[20] Baigorri, R., Fuentes, M., González-Gaitano, G., García-Mina, J.M., Almendros, G., González-Vila, F.J., 2009. Complementary multianalytical approach to study the distinctive structural features of the main humic fractions in solution: Gray humic acid, brown humic acid, and fulvic acid. Journal of Agricultural and Food Chemistry 57, 3266-3272. https://doi.org/10.1021/jf8035353

[21] MacCarthy, P., 2001. The principles of humic substances. Soil Science 166, 738-751. https://doi.org/10.1097/00010694-200111000-00003

[22] Yamauchi, M., Katayama, S., Todoroki, T., Watanable, T., 1984. Total synthesis of fulvic acid. Journal of the Chemical Society, Chemical Communications 1565. https://doi.org/10.1039/c39840001565

[23] Nebbioso, A., Piccolo, A., 2012. Advances in humeomics: Enhanced structural identification of humic molecules after size fractionation of a soil humic acid. Analytica Chimica Acta 720, 77-90. https://doi.org/10.1016/j.aca.2012.01.027

[24] Oliver, B.G., 1983. Dihalo acetonitriles in drinking water: algae and fulvic acid as precursors. Environmental Science & Technology 17, 80-83. https://doi.org/10.1021/es00108a003

[25] Peters, R.J.B., Leer, E.W.B., Galan, L., 1990. Dihaloacetonitriles in Dutch drinking waters. Water Research 24, 797-800. https://doi.org/10.1016/0043-1354(90)90038-8

[26] De Julio, M., De Julio, T.S., Di Bernardo, L., 2013. Influence of the apparent molecular size of humic substances on the efficiency of coagulation using Fenton's reagent. Anais da Academia Brasileira de Ciências 85, 833-848. https://doi.org/10.1590/S0001-37652013005000030

[27] Miao, H., Tao, W., 2008. Ozonation of humic acid in water. Journal of Chemical Technology & Biotechnology 83, 336-344. https://doi.org/10.1002/jctb.1816

[28] Gaffney, J.S., Marley, N.A., Clark, S.B., 1996. Humic and Fulvic Acids and Organic Colloidal Materials in the Environment. pp. 2-16. https://doi.org/10.1021/bk-1996-0651.ch001

[29] Wu, Y., Zhou, S., Ye, X., Zhao, R., Chen, D., 2011. Oxidation and coagulation removal of humic acid using Fenton process. Colloids and Surfaces A: Physicochemical and Engineering Aspects 379, 151-156. https://doi.org/10.1016/j.colsurfa.2010.11.057

[30] Thurman, E.M., Malcolm, R.L., 1981. Preparative isolation of aquatic humic substances. Environmental Science & Technology 15, 463-466. https://doi.org/10.1021/es00086a012

[31] Urbanowska, A., Kabsch-Korbutowicz, M., 2018. Isolation and fractionation of humic substances present in water with the use of anion-exchange resins and ultrafiltration.

Brazilian Journal of Chemical Engineering 35, 1211-1217. https://doi.org/10.1590/0104-6632.20180354s20170514

[32] Riley, J.P., Taylor, D., 1969. The analytical concentration of traces of dissolved organic materials from sea water with Amberlite XAD-1 resin. Analytica Chimica Acta 46, 307-309. https://doi.org/10.1016/S0003-2670(01)95630-2

[33] Stuermer, D.H., Harvey, G.R., 1977. The isolation of humic substances and alcohol-soluble organic matter from seawater. Deep Sea Research 24, 303-309. https://doi.org/10.1016/S0146-6291(77)80010-6

[34] Tsuda, K., Takata, A., Shirai, H., Kozaki, K., Fujitake, N., 2012. A method for quantitative analysis of aquatic humic substances in clear water based on carbon concentration. Analytical Sciences 28, 1017-1020. https://doi.org/10.2116/analsci.28.1017

[35] Kida, M., Maki, K., Takata, A., Kato, T., Tsuda, K., Hayakawa, K., Sugiyama, Y., Fujitake, N., 2015. Quantitative monitoring of aquatic humic substances in Lake Biwa, Japan, using the DAX-8 batch method based on carbon concentrations. Organic Geochemistry 83-84, 153-157. https://doi.org/10.1016/j.orggeochem.2015.03.015

[36] Mahvi, A.H., Maleki, A., Rezaee, R., Safari, M., 2009. Reduction of humic substances in water by application of ultrasound waves and ultraviolet irradiation. Iranian Journal of Environmental Health Science and Engineering 6, 233-240.

[37] Tadini, A.M., Moreira, A.B., Bisinoti, M.C., 2014. Fractionation of aquatic humic substances and dynamic of chromium species in an aquatic body influenced by sugarcane cultivation. Journal of the Brazilian Chemical Society 25, 119-125. https://doi.org/10.5935/0103-5053.20130277

[38] Gu, N., Liu, J., Ye, J., Chang, N., Li, Y.Y., 2019. Bioenergy, ammonia and humic substances recovery from municipal solid waste leachate: A review and process integration. Bioresource Technology 293. https://doi.org/10.1016/j.biortech.2019.122159

[39] Nardi, S., Pizzeghello, D., Muscolo, A., Vianello, A., 2002. Physiological effects of humic substances on higher plants. Soil Biology and Biochemistry 34, 1527-1536. https://doi.org/10.1016/S0038-0717(02)00174-8

[40] Leite, J.M., Pitumpe Arachchige, P.S., Ciampitti, I.A., Hettiarachchi, G.M., Maurmann, L., Trivelin, P.C.O., Prasad, P.V.V., Sunoj, S.V.J., 2020. Co-addition of humic substances and humic acids with urea enhances foliar nitrogen use efficiency in sugarcane (*Saccharum officinarum* L.). Heliyon 6. https://doi.org/10.1016/j.heliyon.2020.e05100

[41] Canellas, L.P., Olivares, F.L., 2014. Physiological responses to humic substances as plant growth promoter. Chemical and Biological Technologies in Agriculture 1, 3. https://doi.org/10.1186/2196-5641-1-3

[42] Zandonadi, D.B., Santos, M.P., Dobbss, L.B., Olivares, F.L., Canellas, L.P., Binzel, M.L., Okorokova-Façanha, A.L., Façanha, A.R., 2010. Nitric oxide mediates humic acids-induced root development and plasma membrane H+-ATPase activation. Planta 231, 1025-1036. https://doi.org/10.1007/s00425-010-1106-0

[43] Zhang, Q., Zou, D., Zeng, X., Li, L., Wang, A., Liu, F., Wang, H., Zeng, Q., Xiao, Z., 2021. Effect of the direct use of biomass in agricultural soil on heavy metals __ activation or immobilization? Environmental Pollution 272, 115989.

https://doi.org/10.1016/j.envpol.2020.115989

[44] Yurishcheva, A.A., Kydralieva, K.A., Zaripova, A.A., Dzhardimalieva, G.I., Pomogaylo, A.D., Jorobekova, S.J., 2013. Sorption of Pb2+ by magnetite coated with humic acids. Journal of Biological Physics and Chemistry 13, 61-68.

[45] Lipczynska-Kochany, E., 2018b. Effect of climate change on humic substances and associated impacts on the quality of surface water and groundwater: A review. Science of The Total Environment 640-641, 1548-1565. https://doi.org/10.1016/j.scitotenv.2018.05.376

[46] Ayekoe, C.Y.P., Robert, D., Lanciné, D.G., 2017. Combination of coagulation-flocculation and heterogeneous photocatalysis for improving the removal of humic substances in real treated water from Agbô River (Ivory-Coast). Catalysis Today 281, 2-13. https://doi.org/10.1016/j.cattod.2016.09.024

[47] Ye, W., Liu, H., Jiang, M., Lin, J., Ye, K., Fang, S., Xu, Y., Zhao, S., Van der Bruggen, B., He, Z., 2019. Sustainable management of landfill leachate concentrate through recovering humic substance as liquid fertilizer by loose nanofiltration. Water Research 157, 555-563. https://doi.org/10.1016/j.watres.2019.02.060

[48] Arancon, N.Q., Edwards, C.A., Lee, S., Byrne, R., 2006. Effects of humic acids from vermicomposts on plant growth. European Journal of Soil Biology 42, S65–S69. https://doi.org/10.1016/j.ejsobi.2006.06.004

[49] Cooper, R.J., Liu, C., Fisher, D.S., 1998. Influence of humic substances on rooting and nutrient content of creeping bentgrass. Crop Science 38, 1639-1644. https://doi.org/10.2135/cropsci1998.0011183X003800060037x

[50] Newcomer, R.W., Nybo, J.P., Newcomer, J.K., 2020. Humate in the upper Cretaceous Fruitland Formation in northwestern New Mexico. New Mexico Geological Society Special Publication 14, 41-46.

[51] McLemore, V.T., 2020. Uranium deposits in the Poison Canyon trend, Ambrosia Lake Subdistrict, Grants Uranium District, McKinley and Cibola Counties, New Mexico. New Mexico Geological Society Special Publication 14, 53-63.

[52] Chotzen, R.A., Polubesova, T., Chefetz, B., Mishael, Y.G., 2016. Adsorption of soil-derived humic acid by seven clay minerals: A systematic study. Clays and Clay Minerals 64, 628-638. https://doi.org/10.1346/CCMN.2016.064027

[53] López-Cervantes, R., Tejo, A.G., Peña-Cervantes, E., Reyes-López, A., Castro-Franco, R., Chávez-González, J.F.J., 2006. Humic substances from different sources on some physical properties of a silty-clay-loamy soil. Terra Latinoamericana 24, 303-309.

[54] Naresh, R.K., Gupta, R.K., Dhaliwal, S.S., Tyagi, K., Tyagi, S., Krishna Prasad, K.S., Jat, L., Tyagi, P., 2018. Clay-humus stability of soil organic matter and microbial biomass under conservation tillage and residue management practices of rice-wheat cropping system: A review. Journal of Pharmacognosy and Phytochemistry 7, 3020-3043.

[55] Miranda, G.A.P., Júnior, J.T.A., Brocchi, E.A., Wang, H., 2021. Humic substances reduce the erodibility of soils in mining areas. Journal of Cleaner Production 279. https://doi.org/10.1016/j.jclepro.2020.123700

[56] Piccolo, A., Spaccini, R., De Martino, A., Scognamiglio, F., di Meo, V., 2019. Soil washing with solutions of humic substances from manure compost

removes heavy metal contaminants as a function of humic molecular composition. Chemosphere 225, 150-156. https://doi.org/10.1016/j.chemosphere.2019.03.019

[57] Bernardino, C.A.R., Mahler, C.F., Preussler, K.H., Novo, L.A.B., 2016. State of the art of phytoremediation in Brazil — Review and perspectives. Water, Air, & Soil Pollution 227, 272. https://doi.org/10.1007/s11270-016-2971-3

[58] Bernardino, C.A.R., Mahler, C.F., Alvarenga, P., Castro, P.M.L., da Silva, E.F., Novo, L.A.B., 2020. Recent advances in phytoremediation of soil contaminated by industrial waste: A road map to a safer environment, in: Bioremediation of Industrial Waste for Environmental Safety. Springer Singapore, Singapore, pp. 207-221. https://doi.org/10.1007/978-981-13-1891-7_10

[59] Patra, D.K., Pradhan, C., Patra, H.K., 2020. Toxic metal decon tamination by phytoremediation appraoch: Concept, challenges, opportunities and future perspectives. Environmental Technology & Innovation. https://doi.org/10.1016/j.eti.2020.100672

Chapter 9

Sewage Sludge to Fertilise Durum Wheat: Effects on Crop and Soil

Pilar Mañas and Jorge de las Heras

Abstract

The vast quantities of degradable waste generated in urban areas may negatively influence the environment if improperly managed. This study examines effects on soil properties, yield and morphological performance of winter wheat (*Triticum turgidum* L. cv. Vitron) after applying composted and air-dried sewage sludge. The experiment was conducted on the field scale in two different farm soil plots Toledo, (central Spain) with different characteristics, especially salinity, concentration of chlorides, sulphates and pH. Three fertiliser treatments were considered: commercial fertiliser; air-dried sewage sludge and composted sewage sludge. Sewage sludge promoted better yields than the commercial fertiliser, and preserved soil physico-chemical characteristics. The sewage sludge application (air-dried and composted) to soil improved the results of the morphological characteristics of the studied wheat in relation to the commercial fertiliser. In the saline soil plot, air-dried sewage sludge improved the morphological characteristics of spikes (length, weight, number of grains per spike), but not final grain weight and, consequently, yield. These were upgraded with the composted sewage sludge. Use of sewage sludge for winter wheat production was the better studied option and proved a sustainable approach to recycle such waste on land.

Keywords: sewage sludge, compost, wheat, spike, soil, reuse, sustainability

1. Introduction

The rising prices of mineral fertilisers have led more farmers to consider organic matter to cover crop nutrient requirements and to maintain soil fertility [1]. It is possible to adequately process organic urban waste and obtain organic fertilisers following aerobic/anaerobic treatments [2]. Aerobic organic waste transformation leads to a well-humified stabilised material (compost) that can be used as fertiliser because it comprises many essential plant nutrients: P, N, Zn, K, Mn, Cu, Fe, etc. Organic matter from biosolids is an important source of essential nutrients for agricultural crop growing [3]. It enhances soil physico-chemical properties, encourages soil activity and microbial growth, and promotes good soil structure which, in turn, improves water holding capacity and aeration. However, this practice poses potential risks related to the accumulation of contaminants in surface soil. Nevertheless, benefits from applying biosolids to farmland have been well documented [4–6], and are fostered by European legislation currently in force; e.g. Council Directive 86/278/EEC on Sewage Sludge [7], which is believed to be the best environmental practice option for most circumstances [8].

Sewage sludge is a by-product from wastewater treatment processes. Given the increases in wastewater treatment plant activity and quantity of produced waste materials, waste management has become a real environmental problem [9]. Fortunately, orientating this waste to agricultural recovery is a management strategy that forms part of sustainable development [10–12]. Its land application is a relevant incentive as regards nutrient recycling, reuse and soil amendment [13, 14], and the fertilising effect has been observed more for well-drained soil [15]. Sewage sludge nutrient content sustains soil fertility, while soil properties are improved by organic constituents. Using sludge in agriculture is one of the solutions that the European Union contemplates for this waste and is considering circular economy practice in which waste becomes a resource.

However, sewage sludge, like other biosolids, can contain large quantities of toxic heavy metals (e.g., Cr, Pb, Ni, Hg and Cd) because industrial wastewater is mixed with sewage. The disagreement about the agricultural application of sewage sludge is related to the quality and safeness of food products [16]. Being sustainable does not mean that it does not incur risks associated with the characteristics of the material itself or with its handling. Those risks must be controlled so they do not have any significant negative effects. The way to do this is to study each case individually because no standard sewage sludge composition exists.

Applying metal-polluted sewage sludge can mean crop damage, soil/water pollution and heavy metal accumulation in the food chain. The magnitude of this problem lies in the composition of sludge and its application rate, crop species and management practices, as in soil properties [14].

Most potential toxicity problems lie in biological toxicity tests, which allow us to observe how a certain toxic or contaminated agent impacts the viability of a living being. Ecotoxicity is calculated by measuring EC_{50}, which is the dose of the compound required for half a population (bacterial in this case) to die. To this end, a leachate is obtained that simulates the transfer of contaminants, which takes place in the medium when it encounters water [17].

Using compost from sewage sludge and organic municipal solid waste in agriculture offers the potential for recycling plant nutrients and to, thus, reduce the employment of mineral fertilisers. Previous studies have confirmed that compost fertilisers enhance microbiological soil physico-chemical properties and enlarge the pool of nutrients, such as P [18] and organic C [14, 19] in soil. Moreover, when composting sewage sludge, health hazards from pathogens (bacteria, protozoa, parasitic helminths, etc.) are reduced [12].

It has been proved than sewage sludge-based fertilisation might be the most effective way to mitigate the negative effects of water stress on wheat yield in arid and semiarid regions [11], and also a good form of substitution towards environmentally friendly agriculture.

Using sewage sludge as a fertiliser has led to promising results in agriculture compared to chemical fertilisers [11, 12, 20].

Moreover, wheat is one of the world's most widely grown crops. In 2019/2020, the European Union produced a wheat volume that came to 153.5 million metric tons [21]. Durum wheat possesses excellent food qualities, namely minerals, gluten and high-fibre content. In Spain, an average of 6 million hectares of cereals is cultivated. It is the sector with the largest territorial base and with nationwide distribution. The 2018 national cereal harvest (marketing year 2018/19) is estimated at 23.26 million tons, 44.45% more than the previous year, and was characterised by a bad season due to drought. In autumn-winter cereals, 28.9% correspond to soft wheat with 6.7 million tons, and 5.7% to durum wheat with 1.32 million tons [22].

Spain could become the 5th community producer with 23.1 Mt. According to the Junta de Comunidades de Castilla-La Mancha (JCCM) [23], wheat production in Castilla-La Mancha (Spain) was 542,000 tons in 2017.

Wheat (*Triticum aestivum* L.) has a prominent place among cereals for its high nutritional value. Its grain contains starch (60–68%), protein (6–21%), fat (1.5–2.0%), cellulose (2.0–2.5%), minerals (1.8%) and vitamins [24]. It is one of the most important grain crops grown all over the world and plays a major role in the economic activity of most countries [25]. Durum Wheat (*Triticum durum*) cultivation is very ancient in semiarid Mediterranean areas [11]. With winter wheat, obtaining high grain yield depends on sufficient numbers of fertile spikes and enough shoot biomass developing [26]. To achieve these objectives, fertiliser composition must be considered and, thus, fertilisers produced from sewage sludge can be a suitable solution.

When we talk about sewage sludge, we usually generalise something that is extremely variable. Sewage sludge is not all the same in composition and danger terms as some hardly contain pollutants and can even be used directly. Others, however, possess many dangerous and polluting components. Our study focused on that produced in a specific place.

The main aim of this field trial was to assess the response of crop and soil that underwent three different fertiliser treatments based on a one-time waste input from a wastewater treatment plant (WWTP). We hypothesised that WWTP waste (composted or air-dried sewage sludge) as the initial fertility driver of soil amendments would enhance soil physico-chemical properties and final durum wheat yields. The study objectives were to investigate the effect of applying: (i) composted sewage sludge, air-dried sewage sludge and commercial fertiliser on the morphological characters of wheat crop (spike and grain) and yield; (ii) the above fertiliser treatments on soil characteristics in relation to the original soil.

2. Materials and methods

2.1 Study area and experimental design

Field trials were run in two dry-land agricultural plots (15 ha each) located in the province of Toledo (central Spain): Villacañas (39°35′17" N; 3°27′45"W) and Quero (39° 33′ 7.16" N; 3° 15′ 37.17" W). These plots have been historically cultivated for winter cereals (oat, barley, rye, or wheat) and sporadically intercalated with fallow. A local weather station (YUTM 4340164, XUTM: 482750, altitude 658 m) reported the mean minimum, mean maximum and mean average temperatures, which were respectively −0.46°C, 25.56°C and 10.93°C. Total precipitation was 275.5 mm.

Three fertiliser treatments were considered: commercial fertiliser (CF); air-dried sewage sludge (SS); composted sewage sludge (CS). We opted to use 15 t ha^{-1} in all three cases because it is the usually applied dose in the local area for durum wheat when using this CF (8–16-8). Considering the composition of SS and CS (**Table 1**), the equivalence in terms N-P-K were SS (6–5-0.3) and CS (4.3–5.3-0.5).

Replication is the most important aspect of an experimental design [27, 28]. Hence each test plot (Villacañas and Quero) was divided into 27 subplots (9 replicates per fertiliser treatment: CF, SS and CS) that were randomly distributed.

Plots were prepared using farm machinery early in December and treatments (CF, SS and SC) were applied to soil. Next 180 kg ha^{-1} of *Triticum turgidum* L. cv. Vitron seeds were sown 2 weeks later. This is a high-yielding durum wheat variety characterised by medium earliness in spike emergence, a light to dark brown speak colour, medium plant height and being well-adapted to all growing regions.

2.2 Characteristics of experimental materials

Before conducting the experiment, the physico-chemical parameters of both air-dried and composted sewage sludge were determined (**Table 1**).

2.2.1 Air-dried sewage sludge

Sewage sludge was obtained from the Alcázar de San Juan WWTP (29,000 inhabitants, central Spain: 391 240 N; 31 120 W. Altitude, 644 m). This WWTP purifies wastewater by a biological activated sludge system. This wastewater is a mixed kind (70% domestic, 30% industrial) and comes from two industrial areas [29] related to winemaking and cheese-making. **Tables 1** and **2** show its chemical characterisation. Organic compounds (PAHs and PCBs), *Salmonella* sp., *E. coli* and ecotoxicity were also determined.

Agronomic Parameters	Air-dried sewage sludge	Composted sewage sludge
pH	7.01	5.830
Moisture (%)	86.46	9.28
Electrical conductivity (μS cm^{-1})	3087	5740
Total organic matter (%)	65.25	52.08
Total N (%)	5.95	4.30
P_2O_5 (%)	6.10	5.47
K_2O (%)	0.31	0.48
CaO (%)	6.78	12.53
MgO (%)	0.87	1.04
Fe (mg kg^{-1} d.m.)	50688	10301
C:N ratio	5.95	6.68

Table 1.
Characteristics of the air-dried and composted sewage sludge. d.m: dry matter.

Parameter (mg kg^{-1})	Soil Villacañas	Soil Quero	R.D. 1310/90 values limit for soil		Air-dried sewage sludge	Composted sewage sludge	R.D. 1310/90 values limit for sewage sludge	
			Soil with pH < 7	Soil with pH > 7			Soil with pH < 7	Soil with pH > 7
Total Cd	0.1	<0.10	1	3.0	0.33	2.38	20	40
Total Cu	<1.00	<1.00	50	210.0	189	116	1	1,75
Total Cr	15.51	2.21	100	150.0	24.6	26.3	300	400
Total Hg	<0.10	<0.10	1	1.5	30.1	29.9	750	1,2
Total Ni	11.37	1.5	30	112.0	512	431	2,5	4
Total Pb	11.30	11.30	50	300.0	0.45	<0.2	16	25
Total Zn	<20.00	<20	150	450.0	30.9	35.1	1	1,5

Table 2.
Metal content in original soils, air-dried sewage sludge and composted sewage sludge.

In view of the analytical results (**Tables 1** and **2**), waste had pH 7.01, a high moisture content (over 80%), showed high electrical conductivity (3,087 μS cm^{-1}) and was rich in organic matter (nearly 70%). The total N (5.95%), P (6.10%) and K (0.31%) contents had an interesting agronomic value because of their fertilising potential. The N content and C:N ratio (5.95) indicated that this waste could be used for composting as a nitrogen source.

No performed analysis detected *Salmonella* sp., although *E. coli* (1.4 10^5 colony-forming units, cfu) was found. According to Spanish Royal Decree 1310/90 [30], metals did not exceed the legal limits (**Table 2**). Otherwise organic compounds are limited by a European Union Directive. According to this legislation, the sum of 16 PAHs (Naphtalene, Acenaphthylene, Acenaphthene, Fluorene, Phenanthrene, Anthracene, Fluorabthene, Pyrene, Benz (a) anthracene, Chrysene, Benzo(b)fluoranthene, Benzo(k) fluoranthene, Benzo(a) pyrene, Indeno, Benzo (g.h.i.) perylene and Dibenz(a.h)anthracene) must to be below 6 mg kg^{-1} dry matter and the sum of the seven PCBs (PCB 28, PCB 52, PCB 101, PCB 118, PCB 138, PCB 153, PCB 180) cannot exceed 0.8 mg kg^{-1} dry matter. The content of PAHs and PCBs in the studied air-dried sewage sludge was less than 10 μg kg^{-1} dry matter and were, therefore, lower than the EU Directive limits.

As the purposes of using sewage sludge was an agriculture application, an ecotoxicity test was carried out to estimate its potential post-application risk [31]. According to Spanish law [32], waste is characterised as ecotoxic and is, therefore, dangerous when the Effective Concentration 50 (EC_{50}) value is 3,000 mg L^{-1}. In view of the analysis results (502,700 mg L^{-1}), it can be stated that the Alcázar de San Juan WWTP sludge is not considered ecotoxic waste.

2.2.2 Composted sewage sludge

The sewage sludge from the Alcázar de San Juan WWTP was composted by a wind-row composting system. Cereal straw (80% dry matter: 4.2% crude protein, 73% fibre (cellulose, hemicellulose and lignin), 0.35% Ca, 0.15% and P, 2.36% K) was used as a carbon source and also for helping to improve the compost structure thanks to its fibrous texture. As the sewage sludge required being dried beforehand, it was spread over an area of the composting plant before being incorporated. The downwards to upwards order of the materials for pile formation was straw plus sludge plus straw. The composting ratio was two parts straw to one part sludge (v/v). The straw and sludge mixture were periodically flipped to homogenise it and to improve the ventilation of piles. Straw was applied without grinding, which not only provided the mixture with C, but also helped to improve the compost structure. The composting process was completed 4 months after starting to dry sewage sludge. **Tables 1** and **2** provide the final composition details.

High values of electrical conductivity (5,740 μS cm^{-1}) and Fe (10,301 mg per kg of dry matter) were detected. CaO (12.53%) and MgO (1.04%) concentrations were higher than that of the air-dried sewage sludge (6.78% and 0.87%, respectively).

Moisture was 9.28%. The C:N ratio after composting was 6.68 (**Table 1**), so it was a fairly mineralised compost. It had high organic matter (52.08%) and total N (4.30%) contents. The concentration of metals (**Table 2**) was lower than that allowed by law. It was not ecotoxic (<2 toxicity units) and was a microbiologically stable compound. Both *Salmonella* sp. and *Escherichia coli* < 1,000 c.f.u g^{-1} were absent.

2.2.3 Commercial fertiliser

NPK 8–16-8 + organic matter (22%) was the commercial fertiliser employed in the control subplots. This fertiliser is frequently used by farmers in the area. Further details from the label are: P_2O_5 (soluble in neutral ammonium citrate and water), 14.50%; P_2O_5 (soluble in water), 11.50%; K_2O (soluble in water), 8.00%; Organic C, 12.50%; Humic acids, 1.00%; Cu, 208 mg kg^{-1} dry matter; Zn, 559 mg kg^{-1} dry matter.

2.3 Sampling and measurement

2.3.1 Soil

Initially a soil sample from each study plot (Villacañas and Quero) was collected at 10 different random points in an "S" pattern at a depth of 0–35 cm with a hand auger. The analysed parameters were EC, Cl^-, SO_4^{2-}, organic matter, N, C:N ratio, P, CO_3^{2-}, K:Mg ratio, Ca:Mg ratio, Fe, Zn, Cu, Mn, B, cation exchange capacity (CEC), K, Na and Ca (**Table 3**). One month after harvesting, a 0–35 cm-deep soil sample was collected from 10 random points of each replicate. Therefore, nine soil samples per treatment (CF, SS and CS) from each test plot (Villacañas and Quero) were collected and analysed. The assessed parameters are presented in **Table 3**.

Parameter	Units	Villacañas	Quero
pH		8.60	9.15
EC	mmhos cm^{-1}	0.52	13.65
Cl^-	mg kg^{-1}	42	153
$SO_4^=$	mg 100^{-1} g	75	2102
Total $CO_3^=$	%	30,5	15.60
Org. matter	(%)	1.69	2.22
Total -N	%	0.09	0.10
Nitric-N	mg kg^{-1}	5.0	19
Soil-N	kg ha^{-1**}	19.23	73
Ass. P	mg kg^{-1}	19	9
Ass. Fe	mg kg^{-1}	2.72	0.35
Ass. Zn	mg kg^{-1}	0.27	0.38
Ass. Cu	mg kg^{-1}	0.78	0.35
Ass. Mn	mg kg^{-1}	8.70	2.15
Ass. B	mg kg^{-1}	0.61	2.45
Ass. K	meq 100^{-1} g	0.66	0.77
Ass. Na	meq 100^{-1} g	1.70	3.80
Ass. Ca	meq 100^{-1} g	20,31	87.14
CEC	meq 100^{-1} g	12.3	10
C:N ratio		11	13
K:Mg ratio		0.60	0.04
Ca:Mg ratio		18.5	4.9

***N-soil was calculated according to the equation N-soil = (N-nitric x 50)/(10 x density), where N-nitric is expressed as mg kg^{-1} and density is 1.*

Table 3.
Characterization of Villacañas and Quero initial soil plots. Ass: assimilable. CEC: Cationic Exchange Capacity.

Each soil sample was air-dried, sieved to <2 mm and analysed by the following techniques: texture by the Bouyoucos hydrometer method; pH by the potentiometric method (saturated soil paste 1:2,5); electrical conductivity; N by the Kjeldahl procedure; extractable P. Soil samples were prepared for the analysis with acid digestion to determine Ca, Mg, Na and K (atomic emission); Fe, Zn, Cu, and Mn by atomic absorption spectroscopy.

2.3.2 Crop

The preharvested random (215 post-sowing days) bunches of plants were cut far away from the bottom, close to soil, at the center of the subplot (replicate) to prevent board effects. Compared to soil sampling, nine sets of collected plants were used (1 sample x 3 treatments x 3 replicates) per treatment. Yield information (kg ha^{-1}) was recorded. Finally, in the laboratory, 25 plants from each set sample were chosen to estimate spike weight, and length, number, and total weight of grains per spike.

2.4 Statistical procedure

The experimental design involved three treatments (CF, SS and CS) and nine replicates per plot. The data about spike weight (g), spike length (cm), number of grains per spike and total weight of the set of grains of that spike (g) were subjected to statistical processing by an analysis of variance. This was done to determine the factor or factors, and the possible interactions between them, which could affect any differences observed among the treatments.

All the statistical calculations were performed with Statgraphic Centurion XV. The ANOVA table decomposes the variability of the studied parameters (spike length, spike weight, number of grains per spike, total weight of grains per spike and yield) in contributions due to several factors. In the analysis, the sum of the type III squares was chosen. Thus, the contribution of each factor was measured by eliminating the effects of the other factors. P-values proved the statistical significance of all the factors, and those less than 0.05 indicated that these factors had a statistically significant effect on each parameter at the 95% confidence level.

3. Results

3.1 Crop

Table 4 displays the effect of the factors plot (Villacañas or Quero), treatment (SS, CS and CF) and plot-treatment interaction on the studied parameters. Spike length, spike weight, and number of grains per spike were significantly influenced by the plot where the crop was grown (Villacañas or Quero), and by treatment (CF, SS or CS). This did not occur with weight of grains per spike and, consequently, with yield. The plot-treatment interaction was not significant in any case.

A simple factorial ANOVA based on the treatment received (**Table 5**) showed that the type of treatment in Villacañas affected all the studied parameters, except for spike weight.

In Quero, only spike length and spike weight were influenced by treatment. No significant differences were observed in either number of grains per spike or weight of grains or yield.

FACTOR	Spike length	Spike weight	Number of grains per spike	Weight of grains per spike	Yield
PLOT (A)	0.0000*	0.0000*	0.0000*	0.0972	0.0972
TREATMENT (B)	0.0000*	0.0131*	0.0329*	0.0594	0.0594
A-B	0.2607	0.1613	0.1568	0.0672	0.0672

**Significant differences.*

Table 4.
P-values for plot (A): Villacañas, Quero; treatment (B): CF, CS, SS; interaction (A)-(B) for average spike length, average spike weight, average number of grains per spike, average total weight of grains per spike and average yield.

3.1.1 Average spike length

In Villacañas (**Table 5**), no significant differences appeared between SS and CS, and these treatments led to a somewhat longer length (5.47 cm and 5.51 cm, respectively) than with CF (5.09 cm).

In Quero (**Table 5**), spike length was longer than that obtained in Villacañas for all the treatments, and was significantly longer than that achieved by applying SS and CS (6.22 cm and 6.12 cm, respectively) in relation to that achieved when applying CF (5.95 cm).

3.1.2 Average spike weight

In Villacañas (**Table 5**), a statistically significant increase in the average spike weight was observed in the plants grown with SS (1.41 g) compared to those grown with CS (1.25 g) or CF (1.28 g).

In Quero (**Table 5**), like spike length, spike weight was also heaiver than in Villacañas. Weight improved with SS (1.68 g) and with CS (1.66 g) compared to the plants grown with CF (1.49 g).

3.1.3 Average number of grains per spike

In Villacañas (**Table 5**), similarly to average spike weight, the average number of grains per spike was SS (20 grains) than in those plots where CF or CS was applied (18 and 17 grains, respectively).

In Quero (**Table 5**), the same occurred for this parameter as in Villacañas. It was in the subplots with SS and CS where the number of grains per spike was significantly bigger (24) than for the other treatment CF (22). Likewise, it was noteworthy that the number of grains per spike in the Quero plot was always bigger than that obtained in the Villacañas plot, where 20 grains per spike were never exceeded.

3.1.4 Average total weight of grains per spike

In Villacañas (**Table 5**), and following the same line as the previously mentioned parameters, the average total weight of grains per spike was significantly heavier (0.84 g) in the subplots with SS than in those where CS (0.72 g) or CF (0.75 g) was applied.

In Quero (**Table 5**), barely any differences were observed with the Villacañas plot, a trend observed of a slightly increased total weight of grains per spike. In this plot with the compost application, a statistically significant heavier average total weight of grains per spike was recorded (0.86 g).

	Treatment	Average spike length (cm)	Average spike weight (g)	Average number of grains per spike	Average total weight of grains per spike (g)	Average yield (kg ha^{-1})
Villacañas	P-value	0.0002*	0.0099	0.0376*	0.0336*	0.0336*
	CF	5.09 ± 0.75a	1.28 ± 0.56a	18 ± 7a	0.75 ± 0.33a	2994 ± 1337ab
	SS	5.47 ± 0.88b	1.41 ± 0.60b	20 ± 7b	0.84 ± 0.37b	3371 ± 1481b
	CS	5.51 ± 0.85b	1.25 ± 0.60a	17 ± 7a	0.72 ± 0.37a	2884 ± 1476 a
Quero	P-value	0.0269*	0.0166*	0.1497	0.1297	0.1297
	CF	5.95 ± 0.70a	1.49 ± 0.50a	22 ± 7a	0.76 ± 0.36a	3056 ± 1437a
	SS	6.22 ± 0.76b	1.68 ± 0.47b	24 ± 7b	0.83 ± 0.33ab	33171 ± 1311ab
	CS	6.12 ± 0.78ab	1.66 ± 0.53b	24 ± 7a	0.86 ± 0.34b	3423 ± 1357b

Significant differences.

Table 5.
P-values and average ± standard deviations of average spike length (cm), average spike weight (g), average number of grains per spike, average total weight of grains per spike (g) and average yield (kg ha^{-1}). Distinct letters mean groups with significant differences at $P < 0.05$ among treatments for the same study plot according to the LSD test.

3.1.5 Yield

The maximum yield was obtained in the Quero plot (3,423 kg ha^{-1}) with the minimum one in Villacañas (2,884 kg ha^{-1}), both cases with CS (**Table 5**). It was, therefore, clear that soil type strongly influenced crop response to the application of the same product. In Villacañas, the highest yield was obtained with SS (3,371 kg ha^{-1}) and the lowest with compost (2,884 kg ha^{-1}). On the contrary, CS treatment gave the highest yield in the Quero plot (3,423 kg ha^{-1}), where the lowest yield was accomplished with CF treatment (3,056 kg ha^{-1}).

3.2 Soil

The initial Villacañas test soil plot (**Table 3**) had a sandy loam texture [33], basic pH (8.6) and was somewhat saline. The contents of Cl^- (42 mg kg^{-1}), CEC (12.3 meq 100^{-1} g), Fe (2.72 mg kg^{-1}), Zn (0.27 mg kg^{-1}) and Cu (0.78 mg kg^{-1}) were medium, while the levels of organic matter (1.69%) and total N (0.09%) were low [34]. Normal N release was indicated by the C:N ratio (11) [35]. Finally, the amounts of P (19 mg kg^{-1}), total CO_3^{2-} (30.5%), K (0.66 meq 100^{-1} g), Na (1.70 meq 100^{-1} g) and Mn (0.70 mg kg^{-1}) were high [34].

Soil from the Quero test plot (**Table 3**) had higher pH (9.15) than the Villacañas one, but the difference lay in a very saline soil. The contents of Cl^- (153 mg kg^{-1}) and K (0.77 meq 100^{-1} g) were high. Na (3.80 meq 100^{-1} g) and Ca (87.14 meq 100^{-1} g) came in large amounts. Organic matter content (2.22%), total CO_3^{2-} (15.63%), CEC (10 meq 100^{-1} g), Zn (0.38 mg kg^{-1}), Cu (0.35 mg kg^{-1}) and Mn (2.15 mg kg^{-1}) were medium [34]. The C:N ratio indicated little N release [35].

The main difference between the original soils of both plots were salinity and SO_4^{2-} content as the Quero plot soil was very saline (EC: 13.65 mmhos cm^{-1}) and rich in gypsum (2,102 mg 100^{-1} g), unlike the Villacañas soil (EC: 0.52 mmhos cm^{-1} and 75 mg 100^{-1} g, respectively).

In the soils to which CF was applied significant differences were observed between Villacañas and Quero for all the analysed parameters (**Table 6**), except for Cu, Mn and CEC.

In the Quero plot soil (**Table 4**), the values of pH (9.14), EC (19.86 mmhos cm^{-1}) and the C:N ratio (6.37), and the contents of Cl^- (286 mg kg^{-1}), SO_4^{2-} (857 mg 100^{-1} g), organic matter (1.6%), N in all its forms (total-N: 0.15%; nitric-N: 25 mg kg^{-1}; soil-N: 94 kg ha^{-1}), Zn (0.33 mg kg^{-1}), B (3.59 mg kg^{-1}), K (1.89 meq 100^{-1} g), Na (6.8 meq 100^{-1}) and Ca (100.12 meq 100^{-1} g) were significantly higher than in the Villacañas plot (pH:8.2; EC: 0.56 mmhos cm^{-1}; C:N ratio: 4.4; Cl^-: 19.33 mg kg^{-1}), $SO_4^{=}$: 64.3 mg 100^{-1} g; organic matter: 0.7%; total-N: 0.09%; nitric-N: 3.67 mg kg^{-1}; soil-N: 14.3 kg ha^{-1}; Zn: 0.19 mg kg^{-1}; B: 0.12 mg kg^{-1}; K: 0.53 meq 100^{-1} g; Na: 0.59 meq 100^{-1} g) and Ca: 25.25 meq 100^{-1} g). On the contrary in Villacañas, the percentages of total CO_3^{2-}(29.45%), the K:Mg (0.56) and Ca:Mg (28.67) ratios, Fe (1.12 mg kg^{-1}) and P (9.3 mg kg^{-1}) contents were significantly higher than those obtained in Quero (CO_3^{2-}: 14.19%; K:Mg and Ca:Mg ratios: 0.06 and 3.03, respectively; Fe: 0.31 mg kg^{-1} and P: < d.l: detection limit).

Regarding SS treatment, no significant differences were observed between Villacañas and Quero with respect to the content of organic matter, total N, total CO_3^{2-}, Cu, Mn or CEC (**Table 6**).

The differences were significantly greater in Quero regarding pH value (9.01), EC (18.56 mmhos cm^{-1}), Cl^- (362.6 mg kg^{-1}), $SO_4^{=}$ (860 mg 100^{-1} g), C:N ratio (6.7), nitric N (38 mg kg^{-1}), soil N (142.6 kg ha^{-1}), Zn (0.92 mg kg^{-1}), B (5.8 mg kg^{-1}), K (2.1 meq 100^{-1} g), Na (5.8 meq 100^{-1} g) and Ca (93.28 meq 100^{-1} g).

		CF			SS			CS		
		P-value	V	Q	P-value	V	Q	P-value	V	Q
pH		0.0133*	8.2 a	9.14 b	0.0238*	8.4 0a	9.01 b	0.0011*	8.4 a	9.01 b
EC	mmhos cm^{-1}	0.0157*	0.56 a	19.86 b	0.0132*	0.36 a	18.56 b	0.0027*	0.47 a	18.38 b
Cl^-	mg kg^{-1}	0.0106*	19.33 a	286 b	0.0082*	19.33 a	362.6 b	0.0006*	20.6 a	327.3 b
$SO_4^=$	mg 100^{-1} g	0.0000*	64.3 a	857 b	0.0000*	46 a	860 b	0.0000*	42.3 a	860 b
Total $CO_3^=$	%	0.0442*	29.45 b	14.19 a	0.1287	39.98 a	15.18 a	0.0386*	32.8 b	13.34 a
Organic matter	(%)	0.0102*	0.7 a	1.6 b	0.1797	0.99 a	1.86 a	0.053	0.8 a	2.25 a
Total -N	%	0.0011*	0.09 a	0.15 b	0.1757	0.13 a	0.16 a	0.0035 a	0.103 a	0.200 b
Nitric-N	mg kg^{-1}	0.0020*	3.67 a	25.0 b	0.0016*	8.0 a	38 b	0.0003*	6.3 a	40.6 b
Soil-N	kg ha^{-1} (**)	0.0021*	14.3 a	94.0 b	0.0018*	30.6 a	142.6 b	0.0003*	24.3 a	152.6 b
Ass. P	mg kg^{-1}	0.0030*	9.3 b	<d.l	0.0016*	12.7 b	<d.l	0.0061*	17.7 b	<d.l
Ass.Fe	mg kg^{-1}	0.0183*	1.12 a	0.31 b	0.1177	2.24 a	0.5 a	0.0005*	2.31 b	0.4 a
Ass.Zn	mg kg^{-1}	0.0087*	0.19 a	0.33 b	0.0681	0.38 a	0.92 b	0.2264	0.68 a	0.93 a
Ass.Cu	mg kg^{-1}	0.0948	0.51 a	0.38 a	0.1265	0.62 a	0.47 a	0.4762	0.57 a	0.49 a
Ass.Mn	mg kg^{-1}	0.4285	4.9 a	3.5 a	0.0717	7.3 a	3.8 a	0.0181*	8.7 b	3.9 a
Ass.B	mg kg^{-1}	0.0004*	0.12 a	3.59 b	0.0458*	0.1 a	5.8 b	0.0055*	0.2 a	4.5 b
CEC	meq 100^{-1} g	0.4398	14.15 a	12.27 a	0.3994	12.9 a	15.2 a	0.1286	13.8 a	15.1 a
Ass.K	meq 100^{-1} g	0.0048*	0.53 a	1.89 b	0.0023*	0.54 a	2.1 b	0.0005*	0.47 a	2.06 b
Ass.Na	meq 100^{-1} g	0.0062*	0.59 a	6.8 b	0.0166*	0.65 a	5.8 b	0.0019*	0.67 a	7.17 b
Ass.Ca	meq 100^{-1} g	0.0001*	25.25 a	100.12 b	0.0001*	26.03 a	93.28 b	0.0000*	24.8 a	100.05 b

	CF			SS			CS		
	P-value	**V**	**Q**	**P-value**	**V**	**Q**	**P-value**	**V**	**Q**
C:N ratio	0.0438*	4.4 a	6.37 b	0.2076	4.3 a	6.7 b	0.1941	4.5 a	6.5 a
K:Mg ratio	0.0003*	0.56 b	0.06 a	0.0006*	0.63 b	0.09 a	0.0032*	0.48 b	0.06 a
Ca:Mg ratio	0.0066*	28.67 b	3.03 a	0.0204*	33.03 b	3.86 a	0.0003*	25.3 b	2.8 a

d.l: detection limit. *Significant differences.*

Table 6.
Comparison of the study plot soils (V: Villacañas; Q: Quero) for each fertiliser treatment. Ass.: assimilable. Distinct letters mean groups with significant differences at $P < 0.05$ between plots for the same treatment according to the LSD test.

In Villacañas, only the values of the K:Mg (0.63) and Ca:Mg (33.03) ratios and Fe (2.24 mg kg^{-1}) and P (12.7 mg kg^{-1}) content were significantly higher.

Finally, after applying CS, there were no significant differences between Villacañas and Quero for organic matter content, C:N ratio, Zn, Cu and CEC.

The differences in Quero were significantly bigger for pH value (9.01), EC (18.32 mmhos cm^{-1}), Cl- (327.3 mg kg^{-1}), SO_4^{2-}(860 mg 100^{-1} g), total N (0.2%), soil N (152.6 kg ha^{-1}), B (4.5 mg kg^{-1}), K, Na and Ca. In Villacañas, only the values of K: Mg (0.48) and Ca:Mg (25.3) ratios, total CO_3^{2-} (32.8%), Fe (2.31 mg kg^{-1}), P (17.7 mg kg^{-1}) and Mn (8.7 mg kg^{-1}) contents were significantly higher.

Thus, we described the results according to the applied treatment by comparing the test plots, but by also comparing to the initial soil, how the application of SS, CS and CF affected Villacañas and Quero soils.

Changes noted in the Villacañas soil after harvest in relation to the initial situation when comparing **Tables 3** and **6**. The differences among treatments at the end of the trial are also shown. Regarding the initial soil, no representative changes were observed in pH, EC, nitric N, total CO_3^{2-}, the K:Mg ratio, assimilable Fe content, Cu, Mn (in SS or CS treatment), CEC and Ca.

Large differences were found in the initial soil for certain parameters. This was the case of the generalised drop in the contents of Cl^-, organic matter, B, K, Na and the C:N ratio. A reduction was also recorded in the SO_4^{2-} content of the plots treated with SS and CS, and in assimilable P and Mn in the soils fertilised with CF.

Conversely, the total N and soil N contents in the plots fertilised with SS and CS increased, as did Zn content.

Finally, when comparing treatments, significant differences in the Villacañas plot soil were observed for the total N percentage and assimilable P content. In both cases when SS and CS were applied, the content of these parameters significantly increased compared to the control (CF).

In general, compared to the Villacañas plot, more changes took place in the Quero soil as regards the initial situation after the trial (see **Tables 3** and **6**). Thus, compared to the initial soil, no representative changes were seen in pH, total CO_3^{2-} or the K:Mg ratio after the applications.

Large differences appeared in the initial soil for most parameters. A generalised decrease in SO_4^{2-} and assimilable P contents, and in the C:N and Ca:Mg ratios, was observed. A slight decrease in organic matter content was found in the subplots treated with CF and SS.

On the contrary for any of the three treatments, EC, Cl^- content, total N, nitric N, Mn, B, K, Na, Ca and the CEC all increased compared to the original soil. Soil N, Fe, Zn and Cu in soil also increased, but only with SS and CS.

Finally, when comparing treatments, significant differences were found in the Quero plot soil for the percentage of total N (increased with CS), and for nitric N, N in soil and Zn (increased with SS and CS).

4. Discussion

Soil is where the root system of plants develops, and it provides them with anchorage, water and nutrients. It is also a non-renewable resource on a human time scale that can be lost through erosion or degraded by contamination. Thus, conservation and improvement should be made priorities.

One of the main problems that affects soil productivity is a rising EC by increasing salts [9]. Applying organic amendments, including sewage sludge, has been widely reported to remediate saline soils and to alleviate salinity and sodicity stress in crops [36, 37]. Our study showed that this did not happen in Villacañas, which coincides with similar studies in which EC did not only not increase, but even lowered [13], which was the case of treatments SS and CS. This could not be stated of the Quero soil. In these subplots, EC increased *vs*. the original soil, especially with CF. This could be a long-term problem because several crops can have problems developing due to high salinity in soil, which would be a limiting factor for future agricultural uses.

CEC expresses the number of positively charged moles that can be exchanged. A soil's change capacity increases with clay and organic matter content because both have electrical negative charges. The major saturating cations of the change complex are, and in this order: Ca^{2+}, Mg^{2+} and K^{+} (Na appears in alkaline soils) [38]. Compared to the original soils (**Table 3**), CEC changed more in Quero than in Villacañas after treatments, especially with SS and CS (**Table 6**). This may act as an asset to decide about using air-dried sewage sludge or composted sewage sludge for fertilising wheat.

The Ca:Mg ratio rose after treatments (**Tables 3** and **6**), including the control CF in Villacañas, but not in Quero. This means that in Villacañas, Ca content in soil was higher than Mg content, which tends to improve soil aeration. In contrast, together with Zn content could be due to, difficulties in Mg absorption in crops. Dalir et al. [39] reported Zn and Ca and Zn and Ni interactions, as well as high Ca concentrations. This was also the case of the studied Villacañas soils, which caused Zn uptake to drastically drop.

Organic matter in soil plays an essential role in the nutrient cycle [9], which refers to a heterogeneous series of compounds whose origin is biological that are found in different degradation states [38]. As expected, the organic matter content of the tested soils lowered after harvest in all cases, except for treatment CS in the Quero plot. However, the general trend of the studies reviewed in the literature was to use fresh sludge applied to soil as an organic amendment, which increases soil organic matter content [9], P and N [13]. As the literature accurately states, what increased was the total post-harvest nitrogen content with treatments SS and CS, especially in the Quero soil. It is noteworthy that organic matter in organic fertilisers has to be mineralised by soil microorganisms so that nutrients are released to soil. This process is slow and depends on not only organic waste characteristics, but also on soil type and the environment (temperature, soil water content, etc.) [40, 41].

In the Villacañas soil, no post-application changes took place in relation to the initial soil in: pH, EC, nitric N, total carbonates, the K:Mg ratio, content in assimilable iron, copper, manganese (with SS or CS), capacity of cationic exchange and calcium. When comparing treatments, the percentage of total nitrogen and assimilable phosphorus content significantly increased after applying SS and CS versus the control.

In Quero, more changes were observed in relation to the initial soil than in Villacañas. When comparing treatments, a significant increase in the percentage of total N was found for SS and CF, but not for CS, while an increase in CF of nitric N, soil N and Zn content occurred with the application of SS and CS.

The presence of sewage sludge in any studied conformation (SS and CS) promoted higher yields than the commercial fertiliser (CF).

The following were observed in the Villacañas test plot: longer spike length with SS and CS than with CF; heavier average spike weight, average number and weight of grains per spike in the plants fertilised with SS; higher yields for the application of SS, like the findings of Zhang et al. [42]; lower yield with CS.

In the Quero test plot, the following were recorded: higher values for spike length, spike weight, number and weight of grains per spike in all the treatments than their equivalents in Villacañas; longer spike length with SS and CS than with

CF; higher values for spike weight and number of grains per spike with SS; heavier weight of grains per spike with CS application, similarly to other studies [43]; better yields with CS; worse yields with CF.

Like other research works [11, 24], if both air-dried and composted sewage sludge are included in soil fertilisation, the studied crop morphological characteristics improved versus the commercial fertiliser.

In any case, yields were lower in the subplots (CF) than in others (SS and CS), and yield potential correlated positively with the aboveground biomass. Thus, better grain yields were obtained in the crops with a bigger accumulated biomass upon maturity [8]. Our results are consistent with other studies [11], and the good nourishment supported by the high concentration of nutrients in sewage sludge positively influenced the growth of the aboveground biomass of wheat. We verified that, depending on the employed product, differences in yield appeared between the highest and lowest figures, which varied between 10% and 15%. Something similar happened in another study, where the sewage sludge application displayed better fertilising capacity than compost, with 12% bigger wheat yields on average [2]. This response could be associated with different chemical N forms [16]. We cannot rule out the possibility that with the high EC obtained for compost, crops could have difficulties to absorb water because of not being able to overcome osmotic pressure to the consequent detriment of nutrient inputs. Nevertheless, we were able to verify that applying sewage sludge as a soil fertiliser improves both durum wheat yield and productivity and better results than mineral-based fertilisers.

The EU Fertilising Product Regulation (FPR) [44], which comes into force as from 2022, must be considered for a future line of work. Above all, limits for a range of contaminants will be considered, such as Cd, which is contained in mineral fertilisers. With the obtained results, the commercialization of composted sewage sludge would only be possible locally. Its agronomic aptitude has been proven, but the study did not determine standardised production. Future studies will have to carry out composting with a system that better controls parameters than that of controlled piles because this method largely depends on environmental conditions as it is located outdoors. Another possibility to improve this study would be to run tests at different sewage sludge/straw concentrations, and with a different straw size to facilitate mixing.

5. Conclusions

In short, the most important conclusions drawn from our study are:

The effect of two amendments constituted by sewage sludge on soil properties and on wheat (*Triticum aestivum* L.) morphological parameters was evaluated.

Next the application of both air-dried and composted sewage sludge to soil improved the results of the studied wheat morphological characteristics in relation to the commercial fertiliser.

In the Villacañas plot, the best wheat results were obtained by applying air-dried sewage sludge to soil.

On the contrary in the Quero plot, where soil was much more saline and pH was higher, air-dried sewage sludge improved the morphological characteristics of spikes (length, weight and number of grains per spike), but not the final grain weight and, consequently, yields. These were upgraded with composted sewage sludge.

Given the composition of the sewage sludge obtained from the Alcázar de San Juan WWTP, we realized that it was an important source of N and P that serves to both provide crops with nutrients and to maintain these elements in the soil to which it is applied. Moreover, as it is not ecotoxic waste, it is suitable for agricultural use and might be an adequate substitute for commercial fertiliser. Precisely because of its nitrogen content, it is an excellent raw material for making compost by balancing the C: N ratio together with other structuring agents, such as straw, pruning remains or any other carbon source.

However, the application of this waste is not completely harmless, and several considerations must be considered:

- The influence of soil salinity on the results as blocking effects or synergies of elements that absorb plants can occur. According to our results, it would seem that it is preferable to use saline soil in a composted form.

- Although the applied dose was not herein evaluated, is essential that future field trials learn what the appropriate product quantity is according to soil type and crop type.

Our findings suggest that it is possible to sustain wheat yields and to maintain soil properties when sludge is used as an organic amendment towards more sustainable fertilisation practices.

Acknowledgements

This project has been funded with support from the Spanish Ministry of Economy and Competitiveness (MINECO). The authors are grateful to the municipally owned company, "Aguas de Alcázar", GESREMAN, and to the Alcázar de San Juan Town Hall, for their financial support through Project: "Aptitud para uso agrícola del agua y lodos procedentes de la depuradora de aguas residuales de Alcázar de San Juan". Authors thank Helen Warburton for reviewing the English.

Author details

Pilar Mañas* and Jorge de las Heras
Higher Technical School of Agricultural and Forestry Engineering, University of Castilla-La Mancha, Albacete, Spain

*Address all correspondence to: mariap.manas@uclm.es

References

[1] Dawson, C.J., Hilton, J., 2011. Fertiliser availability in a resource limited world: Production and recycling of nitrogen and phosphorus. Food Policy 36(Suppl. 1):14-22. doi:10.1016/j.foodpol.2010.11.012.

[2] Debiase, G., Montemurro, F., Fiore, A., Rotolo, C., Farrag, K., Miccolis, A., Brunetti, G., 2016. Organic amendment and minimum tillage in Winter wheat grown in Mediterranean conditions: Effects on yield performance, soil fertility and environmental impact. European Journal of Agronomy 75, 149-157.

[3] Wen, B., Li, L., Zhang, H., Ma, Y., Shan, X., Zhang, S., 2014. Field study on the uptake and translocation of perfluoroalkyl acids (PFAAs) by wheat (Triticum aestivum L.) grown in biosolids-amended soils. Environmental Pollution 184, 547-554.

[4] Esperschuetz, J., Bulman, S., Anderson, C., Lense, O., Horswell, J., Dickinson, N., Robinson, B.H., 2016. 1. Production of biomass crops using biowastes on low-fertility soil: 2. Effect of biowastes on nitrogen transformation processes. Journal of Environmental Quality, 45 (6) 1970-1978.

[5] Withers, P.J.A., Flynn, N.J., Warren, G.P., Taylor, M., Chambers, B.J., 2016. Sustainable management of biosolid phosphorus: a field study. Soil use and management 32 (1) 54-63.

[6] Rigby, H., Smith, S.R., 2014. The nitrogen fertiliser value and other agronomic benefits of industrial biowastes. Nutr. Cycl. Agroecosyst. 98(2):137-154. doi:10.1007/s10705-014-9602-4

[7] Council Directive 86/278/EEC of 12 June 1986 on the protection of the environment, and in particular of the soil, when sewage sludge is used in agriculture

[8] Antille, D.L., Godwin, R.J., Sakrabani, R., Seneweera, S., Tyrrel, S.F., Johnston, A.E., 2017. Field-scale evaluation of biosolids-derived organomineral fertilisers applied to winter wheat in England. Soil fertility and crop nutrition 109(2), 1-21.

[9] Eid, E.M., Alrumman, S.A., El-Bebany, A.F., Fawy, K.F., Taher, M.A., Hesham, A.E., El-Shaboury, G.A., Ahmed, M.T. 2019. Evaluation of the potential of sewage sludge as a valuable fertilizer for wheat (Triticum aestivum L.) crops Environmental Science and Pollution Research, 26:392-401 https://doi.org/10.1007/s11356-018-3617-3

[10] Mosquera-Losada, M.R., Rigueiro-Rodríguez, A., Ferreiro-Domínguez, N., Hopkins, A., Collins, R.P., Fraser, M.D. 2014. Sewage sludge fertilization effects on Q. rubra and pasture production and flora biodiversity. In: EGF at 50: The future of European grasslands. Proceedings of the 25th General Meeting of the European Grassland Federation, Aberystwyth, Wales, 7-11

[11] Boudjabi, S., Kribaa, M., Chenchoun, H. 2019. Sewage sludge fertilization alleviates drought stress and improves physiological adaptation and yield performances in Durum Wheat (Triticum durum): A double-edged sword. Journal of King Saud University – Science 31, 336-344.

[12] Iqbal, Z.K., Ahmad, K., Zameer, F., Mehmood, N., Wajid, K., Iram, A., Nadeem, M., Akhter, P., Ghazzal, M., Akhtar, M., Bashir, H., Ashfaq, A., Mnuir, M., Saleem, I. 2020. The potentiality of toxic metals to wheat crop as affected by municipalsolid waste: An implication to human health. Pure and Applied Biology . Vol. 9, Issue 1, pp809-820. http://dx.doi.org/10.19045/bspab.2020.90087

[13] Bai, Y., Zang, C., Gu, M., Shao, H., Guan, Y., Wang, X., Zhou, X.,

Shan, Y., Feng, K., 2017. Sewage sludge as an initial fertility driver for rapid improvement of mudflat salt-soils. Science of the total Environment 578, 47-55.

[14] Latare, A.M., Kumar, O., Singh, S.K., Gupta, A., 2014. Direct and residual effect of sewage sludge on yield, heavy metals content and soil fertility under rice-wheat system. Ecological engineering 69, 17-24.

[15] Marron, N. 2015. Agronomic and environmental effects of land application of residues in short-rotation tree plantations: A literature review. Biomass and bioenergy 81, 378-400.

[16] Nkoa, R. 2014. Agricultural benefits and environmental risks of soil fertilization with anaerobic digestates: a review. Agron. Sustainable Dev. 34 (2), 473-492.

[17] Sobrero, M.C., Ronco, A. 2004. Ensayo de toxicidad aguda con semillas de lechuga (Lactuca sativa L.). p: 71-79. En: Ensayos Toxicológicos y Métodos de Evaluación de Calidad de Aguas, G. Castillo, Ed., Ottawa, Canadá.

[18] Rehman, R.A., Farooq, M. 2020. Co-composts of sewage sludge, farm manure and rock phosphate can substitute phosphorus fertilizers in rice-wheat cropping system. Journal of Environmental Management 259, 109700. https://doi.org/10.1016/j.jenvman.2019.109700

[19] Hernández, T., Chocano, C., Moreno, J.L., García., C., 2016 Use of compost as an alternative to conventional inorganic fertilisers in intensive lettuce (Lactuca sativa L.) crops. Effects on soil plant. Soil and tillage research 160, 14-22.

[20] Pascual, I.M., Carmen, A., Carlos, G., Polo, A., Sánchez, M. 2007. Effect of water deficit on microbial characteristics in soil amended with sewage sludge or inorganic fertilizer under laboratory conditions. Bioresour. Technol. 98, 29-37.

[21] Shahbandeh, M. 2020. https://www.statista.com/statistics/237912/global-top-wheat-producing-countries/

[22] MAPA. 2020. https://www.mapa.gob.es/en/agricultura/temas/producciones-agricolas/cultivos-herbaceos/cereales/ (accessed on 17 May 2020)

[23] JCCM. 2020. http://www.ies.jccm.es/estadisticas/por-tema/sectores-economicos/agricultura (accessed on 17 May 2020).

[24] Gopinath, M., Kishor, P. 2019. Effect of organic and inorganic fertilizers on yield of wheat crop of *Triticum aestivum* L. Plant Archives Vol. 19 No. 1, pp. 122-124

[25] Ahmed, H.A., Mahdy, A.Y. 2019. Influence of Sowing Dates, Nitrogen Fertilizer under Low Rates and Foliar with Boron on Production and Quality of Wheat. J. Plant Production, Mansoura Univ., Vol. 10 (2): 81-91.

[26] Chen, X.X., Zhang, W., Liang, X.Y., Liu, Y.M.,1, Xu, S.J., Zhao, Q.Y., Du, Y.F.,1, Zhang, L., Chen, X.P., Zou, C.Q. 2019. Physiological and developmental traits associated with the grain yield of winter wheat as affected by phosphorus fertilizer management. Scientific Reports 9:16580 | https://doi.org/10.1038/s41598-019-53000-z

[27] Brundrett, M., Bougher, N., Dell, B., Grove, T., Malajczuk, N. 1996. Working with mycorrhizas in forestry and agriculture. Australian Centre for International Agricultural Research. pp: 374.

[28] Van Frank, G., Goldringer, I., Rivière, P., David, O. 2019. Influence of experimental design on decentralized, on-farm evaluation of populations:

a simulation study. Euphytica 215: 126. https://doi.org/10.1007/s10681-019-2447-9

[29] Mañas, P., Castro, E., De las Heras, J., 2014. Application of treated wastewater and digested sewage sludge to obtain biomass from Cynara cardunculus L. Journal of Cleaner Production 67, 72-78.

[30] RD 1310/1990 (1990) Royal Decree 1310/1990, of 29 October 1990, concerning the correct use of sewage sludge in agriculture.

[31] Phoungthong, K., Zhang, H., Shao, L.M., He, P.J. 2016. Variation of the phytotoxicity of municipal solid waste incinerator bottom ash on wheat (Triticum aestivum L.) seed germination with leaching conditions. Chemosphere 146: 547-554.

[32] BOE. 1989. Orden de 13 de octubre de 1989 por la que se determinan los métodos de caracterización de los residuos tóxicos y peligrosos. https://www.boe.es/eli/es/o/1989/10/13/(5) (accessed on 16 April 2020)

[33] USDA, 1987. Module 3-Textural Soil Classification. Study Guide. United Department of Agriculture. Soil Conservation Service. https://www.wcc.nrcs.usda.gov/ftpref/wntsc/H&H/training/soilMod1/soil-unified-classEntire.pdf

[34] Yañez, J. Análisis de suelos y su interpretación. Horticultura 1989, 49, 75-89.

[35] Quéméner, J., 1985. L'interpretacion des analyzes. Cultivar 22, 107-117.

[36] Ding, Z., Kheir, M.S.A., Ali, M.G.M.,2, Osama A. M. Ali, Aly O.A.M., Abdelaal, A.I.N., Lin, X., Zhou, Z., Wang, B., Liu, B., He, Z. 2020 The integrated effect of salinity, organic amendments, phosphorus fertilizers, and deficit irrigation on soil properties, phosphorus fractionation and wheat productivity. Scientific Reports, 10:2736 | https://doi.org/10.1038/s41598-020-59650-8

[37] Seleiman, M.F., Kheir, A.M.S. 2018. Saline soil properties, quality and productivity of wheat grown with bagasse ash and thiourea in different climatic zones. Chemosphere 193, 538-546.

[38] Amorós, J.A., Bravo, S., Pérez, C., García, F.J., Jiménez, R., Sánchez, M., 2017. El suelo y su influencia en el cultivo de la vid. Agricultura (1003): 114-119.

[39] Dalir, N., Tandy, S., Gramlich, A., Khoshgoftarmanesh, A., Schulin, R., 2017. Efffects of nickel on zinc uptake and translocation in two wheat cultivars differing in zinc efficiency. Environmental and experimental botany 134, 96-101.

[40] Zech, W., Senesi, N., Guggenberger, G., Kaiser, K., Lehmann, J., Miano, T.M., Miltner, A., Schroth, G., 1997. Factors controlling humification and mineralization of soil organic matter in the tropics. Geoderma 79, 117-161.

[41] Hirzel J., Salazar, F. 2011. Uso de Enmiendas Orgánicas como Fuente de Fertilización en Cultivos. Fertilización de Cultivos en Chile. In: Hirzel, J., Ed., Colección Libros INIA No. 28, Chillán.

[42] Zhang, Y., Chen, T., Liao, Y., Reid, B.J., Chi, H., Hou, Y., Cai, C. 2016. Modest amendment of sewage sludge biochar to reduce the accumulation of cadmium into rice (Oryza sativa L.): a field study. Environ. Pollut. 216, 819-825. https://doi.org/10.1016/j.envpol.2016.06.053.

[43] Lakhdar, A., Iannelli, M.A., Debez, A., Massacci, A., Jedidi, N., Abdelly, C. 2010. Effect of municipal solid waste compost and sewage sludge use on wheat (Triticum durum): growth, heavy

metal accumulation, and antioxidant activity. J. Sci. Food Agric. 90, 965-971. https://doi.org/10.1002/jsfa.3904.

[44] FPR; 2019. https://www.consilium.europa.eu/en/press/press-releases/2019/05/21/eu-adopts-new-rules-on-fertilisers/